Plutonium:
Deadly Gold of the
Nuclear Age

Contributing authors (alphabetically):

Alexandra Brooks, IEER
Bernd Franke, IEER
Milton Hoenig, IEER
Howard Hu, IPPNW
Arjun Makhijani, IEER
Scott Saleska, IEER
Katherine Yih, IPPNW

Reviewers (alphabetically):

Thomas Cochran, Natural Resources Defense Council, USA
Charles Forsberg, Oak Ridge National Laboratory, USA
Rebecca Johnson, Greenpeace UK
John H. Large, Large & Associates Consulting Engineers, UK
Robert Morgan, Washington University in St. Louis, USA
Alan Phillips, Canadian Physicians for the Prevention of Nuclear War
Joseph Rotblat, Pugwash Conferences on Science and World Affairs
Kathleen M. Tucker, Health & Energy Institute, USA
Frank von Hippel, Princeton University, USA

Book and cover design:

Nick Thorkelson, IPPNW

Financial support:

W. Alton Jones Foundation
John D. and Catherine T. MacArthur Foundation
New-Land Foundation
Simons Foundation
Svenska Läkare Mot Kärnvapen
(Swedish Physicians Against Nuclear Weapons)

PLUTONIUM
Deadly Gold of the Nuclear Age

by a special commission of
International Physicians for the Prevention
of Nuclear War
and
The Institute for
Energy and Environmental Research

Prepared under the direction of:

Howard Hu, M.D., M.P.H., Sc.D.
Director, IPPNW Commission

Arjun Makhijani, Ph.D.
President, IEER

Katherine Yih, Ph.D.
Coordinator, IPPNW Commission

International Physicians Press
Cambridge, Massachusetts

International Physicians for the Prevention of Nuclear War
126 Rogers Street
Cambridge, Massachusetts 02142-1096, USA
Telephone: (617)868-5050
Fax: (617)868-2560

Institute for Energy and Environmental Research
6935 Laurel Avenue
Takoma Park, Maryland 20912, USA
Telephone: (301)270-5500
Fax: (301)270-3029

ISBN 0-9634455-0-2

Library of Congress Catalog Card Number: 92-74930

Errata

Plutonium: Deadly Gold of the Nuclear Age

Please note the following changes for *Plutonium: Deadly Gold of the Nuclear Age*, International Physicians Press, Cambridge, Mass., 1992. Some of these changes are corrections, while others supply more recent information. The changed words or phrases are underlined.

4: The third line from the bottom should explain that plutonium has <u>15 isotopes</u>, not 13.

37: The first line of the second paragraph under China should read: replace the word Lanzhou with <u>Guangyuan</u>

43: Table 2.3, the figures for La Hague, France should read: <u>Belgium 1.17; France 15.7; Germany 14.58; Japan 1.17; Netherlands 0.67; Switzerland 1.11, for a total of 34.4 metric tons</u>.

43: Table 2.3, the figures for Marcoule should read: <u>France 5.6 metric tons; Spain zero</u> (plutonium from the Vandellos reactor in Spain belongs to France). The total for civilian plutonium from Marcoule is <u>5.6</u> metric tons.

43: The date on the last line in the notes should be <u>1991</u>.

69: Second to last line from the bottom, the plutonium production figures for Marcoule, should be <u>11.6 tons, including an estimated 6 tons of plutonium for military purposes</u>.

71: Table 3.3, under heading "Sr-90 + Cs-137 curies" for La Hague should <u>204 million</u>, and for Marcoule should be <u>70 million</u> curies.

75: In the first line of the third paragraph, the figure 223 million was an error in the English edition. It should be changed to <u>22.3 million</u>. This does not include wastes discharged to Lake Karachay.

85: Second to last line, change 1 kilogram to <u>10 kilograms</u>.

For new plutonium data refer to: Albright, D., Berkhout, F. and Walker, F. 1993. *World Inventory of Plutonium and Highly Enriched Uranium*. A SIPRI book. New York: Oxford University Press.

Table of Contents

Chapter 3
RADIOACTIVE WASTES FROM PLUTONIUM
PRODUCTION

List of Figures

List of Tables

Preface

If someone tries to creep past your house with a ticking time bomb, naturally there is reason for concern.

Spokesperson for the Foreign Ministry of Argentina on plans to transport plutonium from France and Britain to Japan by ship, 1 October 1992

The IAEA has told Japan that its plans to store huge quantities of plutonium here for its ambitious civilian nuclear program could pose "political and security problems" in Asia. . . . officials . . . are chiefly worried that other nations with nuclear ambitions, including North Korea and Taiwan, could use the Japanese precedent to insist that they, too, should have nuclear reprocessing installations and plutonium stockpiles.

Press report from Tokyo, 13 April 1992

Over half the Federal facilities are not in compliance.

Senator George J. Mitchell (D-Maine) on U.S. Department of Energy nuclear installations and Federal waste disposal laws, 23 Sept. 1992

Kadhim and colleagues describe experiments which, unexpectedly, show that alpha-particle irradiation of a cell may induce a transmitted genetic instability that may well result in . . . genetic abnormalities later. . . . [They] point to the possible importance of their findings in radiation-induced leukæmogenesis and childhood leukæmia clusters associated with nuclear sites.

H. John Evans, commenting on findings published in Nature, 20 February 1992

There is lots of cancer, and many people have radiation poisoning. Children are being born with sicknesses that are related to the accident or other past events.

Yadim Nikitin, Director of Social Welfare for the Chelyabinsk region, where Soviet nuclear weapons were made, 21 Feb. 1992

I N THE YEAR after the Berlin Wall came tumbling down, the USSR crumbled, and the doomsday clock reversed itself, the nightmare of a world poised on the brink of nuclear Armaggedon might seem to be receding.

It is to be hoped. But these recent press reports provide a snapshot of a world in which, as superpower tension dissipates, we are discovering

islands of formidable destruction and danger and new problems for generations to come — the legacy of decades of nuclear weapons production and
testing. Without a single nuclear weapon being detonated in conflict since
the end of World War II, thousands of square miles of land have become
contaminated, ecosystems have been polluted and disturbed, and many
thousands of people have been exposed to radiation, often with tragic consequences. To make matters worse, the threat of nuclear weapons proliferation is rising, at a time when regional conflicts have dramatically increased.

What exactly is the price the world has had to pay? What new problems have arisen, and what are the risks inherited by future generations?
These are the questions that led International Physicians for the Prevention of Nuclear War (IPPNW) to create a research commission in December of 1988. The objective of the Commission has been to describe the
health and environmental effects of nuclear weapons production and testing in scientific yet accessible terms, in order to provide the public with
some understanding of the true costs of merely building and testing
nuclear weapons.

In 1991, under the direction of Dr. Anthony Robbins, the Commission
published *Radioactive Heaven and Earth.* In this remarkable report, the
Commission marshalled the expertise of physicians and other professionals from around the world and focused its investigation on the health and
environmental effects of atmospheric and underground nuclear weapons
testing. Among the original contributions it made were more refined estimates of the number of cancer cases and deaths to be expected from global scattering of fallout, a country-by-country review of nuclear weapons
testing programs, and calculations of the inventories of radionuclides left
underground by underground testing.

In the present report, we have moved our attention to plutonium, the
very heart of nuclear weaponry. One of the most toxic substances known,
plutonium continues to emit radiation for tens of thousands of years. We
summarize new information on the toxicity of plutonium; calculate the
amounts of plutonium currently held by the five nuclear weapons powers;
detail the connection between military and civilian uses of plutonium;
review the quantities and types of highly radioactive liquid wastes generated during plutonium production; analyze the risks posed by methods of
storage and disposal of these wastes; and review new information on the
health and environmental damage resulting from accidents involving plutonium and radioactive liquid wastes, including the 1957 waste tank explosion at the Chelyabinsk-65 weapons production site in the Soviet Union.

As with *Radioactive Heaven and Earth,* a physician will find many of the revelations in this report quite disturbing. Plutonium, exposure to which remains difficult to detect, may be even more carcinogenic than previously thought. Large-scale dumping of high-level radioactive wastes from plutonium production occurred in the Soviet Union. The potential for high-level radioactive wastes to explode exists wherever they have been stored in tanks, including in the U.S. No operational method for disposal yet exists.

With this knowledge, current events take on new meaning. Is it wise for Japan to import plutonium for its civilian reactor program? Are there risks to the continued production of plutonium at Chelyabinsk, where it is estimated that plutonium is still accumulating at the rate of 2.5 tons per year? Should the dismantling process of nuclear weapons and supervision of plutonium stocks be put under international control?

We must note that official secrecy still blocks public access to information with which to ponder such questions. It is unjustifiable that governments continue to withhold information vital to environmental protection and human health.

In spite of the obstacles, this study is the first to provide a global accounting of the risks and potential hazards of plutonium production; we prepared it in the hopes that the public would use it to address — and demand answers to — questions regarding plutonium production, use, and disposal.

Howard Hu, M.D., M.P.H., Sc.D.

Director of the IPPNW Commission

Assistant Professor
Harvard School of Public Health
Harvard Medical School
Associate Physician
Brigham and Women's Hospital

Boston, USA
9 October 1992

Acknowledgements

Many individuals and institutions contributed to the research, writing, review, editing, and production of this book, to whom the authors are most grateful.

We would like to give special recognition to the insight of Robert Alvarez, whose FOIA request a decade ago led to the first public disclosure of explosion dangers in U.S. high-level waste tanks.

We also wish to thank Anthony Robbins, former Director of the IPPNW Commission, for his direction and vision in the original conceptualization of the book. He has worked to publicize the dangers of high-level waste from plutonium production in the U.S. since even before his involvement with the Commission.

We acknowledge the related work and moral support of the Physicians Task Force on the Health Risks of Nuclear Weapons Production, which U.S. Physicians for Social Responsibility established at the same time IPPNW started the Commission. Their recently published investigation of the epidemiological practices of the U.S. Department of Energy complements our own research and is cited later in these pages.

Our reviewers were, without exception, conscientious and thorough in their work; many made themselves available to answer our questions long after submitting written critiques. They are: Thomas Cochran, Charles Forsberg, Rebecca Johnson, John H. Large, Robert Morgan, Alan Phillips, Joseph Rotblat, Kathleen M. Tucker, and Frank von Hippel. (The directors of the project accept full responsibility for any errors, however, as well as for the point of view of the book and its recommendations.) Alan Phillips, moreover, wrote the glossary and the first draft of the appendix on nuclear physics.

Rebecca Johnson deserves special acknowledgement and thanks for having commissioned invaluable overviews of nuclear weapons production in the U.K. (Peden 1991) and in France (Barrillot 1991).

In addition, we gratefully recognize the following people, who provided crucial information: David Albright, Marina Degteva, Steven Dolley, Marcia Goldberg, Mira Kossenko, Thomas Marshall, Todd Martin, Leroy Moore, James Phillips, Lydia Popova, Pete Roche, Monique Sené, and Jim Thomas.

We also thank Lois Chalmers for bibliographic searches, Freda Hur for administrative support, Lynn Martin for photo research, Alexandra Taylor and Frederick Walters for library research, and Soviet/Russian IPPNW for arranging and translating interviews with a Soviet official.

Finally, we applaud the art of Nick Thorkelson, who designed the book, and of Robert Del Tredici, who provided many of the photographs.

Chapter 1
Nuclear Gold or Nuclear Poison?

Gold is a wonderful thing! Whoever owns it is lord of all he wants. With gold it is even possible to open for souls the way to paradise!

—Christopher Columbus, 1503 [1]

[I]f the problem of the proper use of this weapon can be solved, we would have the opportunity to bring the world into a pattern in which the peace of the world and our civilization can be saved.

—Henry Stimson, Secretary of War, 1945 [2]

Nuclear Gold

L UST FOR GOLD inspired the conquistadores who invaded the Americas. Plutonium[3] has put the same messianic gleam in the eyes of both those who have wanted to control the world through weaponry and those who have believed it would provide an unlimited source of energy.

Plutonium, like gold, was to be the currency of power and wealth throughout the world. And like gold, it has wreaked environmental damage and endangered the lives of people around the globe.[4]

---

1 Columbus, quoted in Wright 1992, p. 11.
2 Stimson 1945.
3 Unless otherwise specified, all references to plutonium (abbreviated "Pu") in this report refer to plutonium-239, the fissile isotope used to make nuclear weapons. Other isotopes of plutonium will have their atomic weights appended — for instance, plutonium-240.
4 See, for instance, IPPNW and IEER 1991 for a description of the role of the lands of tribal and colonized peoples in the history of nuclear weapons testing throughout the world.

Figure 1.1. Replicas of "Little Boy," the uranium bomb dropped on Hiroshima, and "Fat Man," the plutonium bomb dropped on Nagasaki; Bradbury Science Museum, Los Alamos, New Mexico. The two bombs together killed or seriously injured about 200,000 people. Photo by Robert Del Tredici.

Unlike gold, plutonium is found in only minute quantities in nature; essentially all the plutonium in the world is of twentieth-century origin and is man-made.[5]

The accumulation of plutonium and other nuclear weapons materials became a primary aim of governmental policies both in the United States and in the Soviet Union in the decades after World War II. Even the definitions of victory in a nuclear war have been put in terms of which side would have more plutonium and nuclear weapons left over afterwards, rather than in terms of human well-being. One commentator put it this way:

> The strategic stability of regime A is based on the fact that both sides are deprived of any incentive ever to strike first. Since it takes roughly two warheads to destroy one enemy silo, an attacker must expend two of his missiles

---

5 It should be noted that about two billion years ago, a quantity of plutonium (long since decayed away) must have been created in at least one known "natural" underground reactor in what is now Gabon, West Africa. This phenomenon was made possible by a high concentration of uranium ore and by the fact that the percentage of fissile uranium-235 so long ago was much higher than the 0.7 percent found in today's uranium ores. (Eisenbud 1987, p. 171.) Minute quantities of plutonium can be measured in uranium ore bodies today.

to destroy one of the enemy's. A first strike disarms the attacker. The aggressor ends up worse off than the aggressed.[6]

Plutonium was also to provide deliverance on the civilian side, though also not without a certain price. Alvin Weinberg, former director of the Oak Ridge National Laboratory, where the chemical process was developed for the first large-scale plutonium production plants, said in 1972, "We nuclear people have made a Faustian bargain with society. On one hand, we offer — in the [plutonium] breeder reactor — an almost inexhaustible source of energy. . . . But the price that we demand of society for this magical energy source is both a vigilance and longevity of our social institutions to which we are quite unaccustomed." (Weinberg 1972.)

Glenn Seaborg, who participated in the isolation of plutonium in 1941,[7] also waxed eloquent about the wonders of this element. According to Daniel Ford, Seaborg had the kind of religious streak that would qualify him as a believing member of the nuclear priesthood:

> The future of civilization, as Seaborg saw it, was in the hands of the nuclear scientists who formed the elite team that would, 'build a new world through technology.'. . .
>
> Seaborg focused his attention. . . on a visionary dream of atomic-powered plenty. According to his prospectus on its possible applications, nuclear energy was a magician's potion that could free industrial society permanently from all practical bounds. Millions of homes could be heated and lighted by a single large nuclear reactor. . . . The deserts could be made to bloom, sea water could be made potable, rivers diverted — all as a result, he prophesied, of 'planetary engineering' made possible by the miraculous new element that he had discovered. . . . There would be nuclear-powered earth-to-moon shuttles, nuclear-powered artificial hearts, plutonium-heated swimsuits for SCUBA divers, and much more. . . . 'My only fear [Seaborg stated] is that I may be underestimating the possibilities.'[8]

As we shall see, the long-term environmental and security legacy of plutonium is quite different from these predictions. Just as the quest for gold left in its wake many human tragedies across the Americas, the quest for plutonium has created many tragedies, some of which we do not know how to address to this very day.

6 Charles Krauthammer, quoted in Cohn 1987, p. 22.

7 The first milligram quantities of plutonium were not created in a reactor, but by the irradiation of uranyl nitrate solution by the cyclotron at the University of California at Berkeley.

8 Ford 1982, pp. 23–24.

Since 1941, when plutonium was created in the United States, huge stocks of it have accumulated in several countries. During World War II and immediately thereafter, the United States was in a great hurry to make plutonium in order to build up its nuclear arsenal. Immediately after World War II, the Soviet Union joined the rush to make plutonium for nuclear weapons. Today, the U.K., France, and China also have nuclear weapons containing plutonium, as does Israel.[9] India exploded a plutonium-containing nuclear device in 1974 and has stocks of plutonium that could be used to make nuclear weapons. Non-nuclear-weapons states also maintain stocks of plutonium for their civilian nuclear power programs. Notable among these for the size of their stocks are Germany and Japan. (The U.K. and France currently hold the greater part of the plutonium stocks owned by Germany and Japan.)

The true security and environmental dimensions of the plutonium problem are only now becoming evident. With the end of the Cold War, the difficulty and expense of dealing with pollution and health problems resulting from plutonium production are beginning to be calculated. Even with great efforts, there is no doubt that many of the environmental and health burdens of plutonium production will be passed on to future generations.

Properties of Plutonium

The Appendix presents some background material on nuclear physics, including discussion of subatomic particles, radiation, radiobiology, and nuclear fission and fusion. Together with the glossary, which defines units of radioactivity and other technical terms, this will be useful to the layperson in understanding the following sections.

Plutonium is one of a class of elements known as "transuranic" elements because they have higher atomic numbers than uranium, which is the last natural element in the periodic table.[10] It has 13 isotopes, ranging in atomic weight from 234 to 246. Plutonium is a silvery metal, whose freshly exposed surface resembles iron or nickel in appearance. Plutonium

---

9 Spector 1988; Hersh 1991.

10 Elements are arranged according to their atomic number in the periodic table. Uranium has an atomic number of 92. The next element in the periodic table is neptunium, atomic number 93. Plutonium is next, with an atomic number of 94. The atomic number of an element is equal to the number of protons in the nucleus. The nominal atomic weight is given by the number of protons plus the number of neutrons in the nucleus.

metal comes in densities ranging from 16 to 20 grams per cubic centimeter (depending on its crystal structure) — about 50 percent more dense than lead — making it one of the densest substances known. It has a melting point of 640 degrees C, and a boiling point of 3,187 degrees C. Plutonium metal is very reactive chemically, and it is readily oxidized in humid air to plutonium dioxide (PuO_2), the most common form of plutonium in the environment.[11]

The plutonium in most nuclear weapons is in the metal (unoxidized) form. In this form, plutonium is pyrophoric, i.e., it is capable of igniting spontaneously when exposed to air. Thus, one of the hazards of plutonium in nuclear weapons production is its pyrophoricity. Further, the burning of plutonium creates fine particles, which are far more biologically hazardous than larger particles, for a given amount of plutonium released (see the section below entitled "Nuclear Poison").

The most common isotope of plutonium, and the main one of interest for nuclear weapons (and nuclear power plants), is plutonium-239. Plutonium-239 is one of the few relatively easily obtainable nuclides that is fissile, that is, capable of sustaining a nuclear chain reaction.[12] In other words, the fission of an atom of plutonium-239 releases a sufficient number of neutrons (on average) to produce a fission of another plutonium-239 atom without any external source of neutrons, provided a sufficient amount (called a "critical mass") of plutonium is available. This ability of plutonium-239 to sustain a nuclear chain reaction is the key to its use in nuclear weapons and for nuclear power production. (The other major fissile material that has been used for both nuclear weapons and power production is uranium-235, the only fissile material that occurs naturally in significant quantity. It takes about four times as much highly enriched uranium to make a nuclear weapon as plutonium-239.)

11 Benedict et al. 1981, pp. 430–431, 448.

12 A "fissile" nuclide is one that is fissionable with slow neutrons, and in which the rate of fissions induced is such that more than one new neutron is produced for each neutron absorbed (which is necessary for the reaction to be self-sustaining). Although all elements beyond lead can be made to fission with neutrons of sufficiently high energy, the only readily available long-lived nuclides that can sustain a fission reaction with slow (low-energy) neutrons are uranium-233, uranium-235, plutonium-239, and plutonium-241. (Benedict et al. 1981, p. 42.) Although plutonium-241, like plutonium-239, is fissile, it is less practical than plutonium-239 for a number of reasons. For example, it is less common than plutonium-239, it is difficult to obtain with high purity, and it decays to produce americium-241, which emits gamma radiation, thereby increasing radiological hazards to workers.

Radiation from Plutonium

An important characteristic of any radioactive material is its half-life. The half-life is a measure of the longevity of an element. Since radioactive materials emit radiation as a result of decay, a given amount of a radioactive material will not last indefinitely but will gradually be transformed into other elements. One half-life is the amount of time it takes for half of a given amount of radioactive material to decay away. In other words, after one half-life has passed, one-half of the original amount remains; after two half-lives, one-fourth remains; after three half-lives, one-eighth — and so on. After ten half-lives have passed, less than one-thousandth of the original amount remains; after 20, less than one-millionth.

The half-lives of the important isotopes of plutonium are as follows:[13]

plutonium-238 — 88 years
plutonium-239 — 24,000 years
plutonium-240 — 6,500 years
plutonium-241 — 14 years

The radioactivity of the various plutonium isotopes per unit weight is in inverse proportion to the half-life of that isotope.[14]

Besides plutonium-239, the only other isotope used in considerable quantities is plutonium-238, which serves as a source of heat for thermoelectric generators used on some satellites and in other applications to provide small amounts of electricity. Plutonium-238 is about 300 times more radioactive than plutonium-239 per unit of weight, since its half-life is about that much shorter.

Plutonium-239 decays by emitting an alpha particle, which contains two protons and two neutrons, like the nucleus of a helium atom. The nucleus that remains after the emission of an alpha particle is also radioactive — it is uranium-235. The decay can be represented as follows:

plutonium-239 —> uranium-235 + alpha particle

13 Diverse sources give slightly different half-lives for various elements and isotopes, including plutonium-239. Here we use data from CRC 1988, rounded to two significant digits, unless otherwise stated.

14 To be precise, the radioactivity per unit weight of a radionuclide is exactly inversely proportional to the product of its half-life and atomic weight. Since the various nuclides of plutonium have approximately equal atomic weights, the effect of differing atomic weights in this case can be neglected.

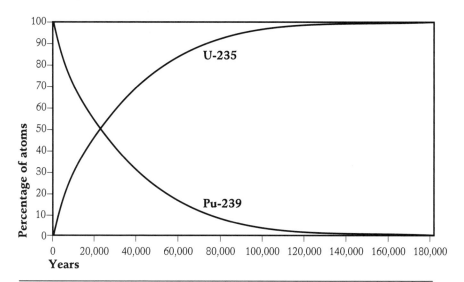

Figure 1.2. Plutonium-239, with a half-life of about 24,000 years, decays to uranium-235. (Uranium-235 decay products are not shown.)

Plutonium-239 decay also results in the release of some gamma radiation.[15]

Thus, as plutonium-239 decays, its inventory or stock decreases and the inventory of uranium-235 correspondingly increases. (Uranium-235 also decays radioactively into other elements, in what is called a decay chain. However, since uranium-235 has a far longer half-life than plutonium-239 (about 700 million years for uranium-235 compared to 24,000 years for plutonium-239), essentially all the plutonium is converted into uranium-235 before any substantial decay of uranium-235 takes place.[16])

15 Most of the modes of plutonium-239 decay leave the resulting uranium-235 nucleus in an excited state. The emission of one or more gamma ray photons (i.e., electromagnetic energy) brings the U-235 nucleus into the ground state. Thus any amount of plutonium-239 will have a small amount of gamma radioactivity as well.

16 Note that the total radioactivity in the uranium-235 that results from plutonium decay is far lower than that in the plutonium. Each atom of plutonium yields one atom of uranium-235. However, since uranium-235 has a half-life about 29,000 times that of plutonium, it decays that many times slower, and so its specific activity (the amount of radioactivity per unit weight) is correspondingly lower. Therefore, one nuclear weapon containing 4 kilograms of plutonium would amount to about 250 curies of radioactivity. Decay of this would yield slightly less than 4 kilograms of uranium-235 (since some of the matter escapes as helium gas — the alpha radiation), but this would have a total radioactivity equal to about 250/29,000, or less than 0.01, curie.

One kilogram of plutonium-239 has a radioactivity level of about 63 curies (or about 2.1 trillion becquerels).

Plutonium-239 is difficult to detect since its gamma radiation is weak and since alpha radiation is rather hard to detect due to its short range. This is especially the case with small quantities of plutonium; nonetheless, such quantities can be lethal.

Nuclear Poison

INTRODUCTION

Plutonium is among the most dangerous of substances. In the initial years after it was discovered, Colonel Stafford L. Warren called it "the most poisonous chemical known."[17] Although substances other than plutonium can produce toxic effects that are more rapidly lethal, Colonel Warren's comment has particular relevance with respect to chromosomal damage and cancer, as we shall see below.

As a metal exposed to the natural environment, plutonium can produce enough heat to boil water and is highly chemically reactive. When in contact with living tissue at high enough levels of exposure, plutonium will cause direct tissue death. Animals experimentally exposed to high concentrations of plutonium by inhalation or injection incur acute damage to the lungs, liver, and hematopoietic (blood-forming) system, and show other manifestations of acute tissue injury.[18] Surviving animals are scarred and develop a number of chronic conditions.[19] Such high level exposure, however, is unlikely to occur to the general public even under a worst-case scenario. Of greatest concern are the radiobiological effects of plutonium, especially cancer, at low levels of exposure.

In the ensuing summary of plutonium toxicity, heavy reliance is made on animal studies, particularly studies of beagle dogs conducted by contractors of the United States Department of Energy (DOE). Investigations of humans exposed solely to plutonium are limited to small case-series studies. Occupational studies of nuclear weapons production workers have provided some data on humans exposed to mixtures of radioactive compounds including plutonium; however, they are few in number and suffer from a number of inaccuracies, omissions, misinterpretations, and other

17 Warren 1946.

18 Thompson 1989.

19 Thompson 1989.

methodological problems. A recent review by a task force of Physicians for Social Responsibility (the United States affiliate of IPPNW) summarized the difficulties posed by the methodological problems as well as the wall of secrecy that surrounded many of these studies in the United States.[20]

CARCINOGENIC MECHANISMS

As a carcinogen, plutonium is dangerous principally because of its alpha (rather than gamma) radiation, and primarily when it is inside the body rather than when outside. When plutonium is in the body, even in small quantities, its alpha radiation causes biological damage. Alpha particles, being heavy, ionize atoms more effectively than electrons and therefore lose their energy and are stopped in a much shorter distance. Because of the relatively many ionizations per unit distance (and per unit of energy lost), alpha radiation is called "high linear energy transfer" radiation ("high LET radiation"), as distinct from the relatively low energy transfer per unit length of photons and electrons ("low LET radiation"). Since alpha particles have a very short range in matter, about 50 micrometers in soft tissue, the energy delivery is more highly concentrated compared to energy from lower LET radiation sources such as beta or gamma radiation emitters. This results in far more biological damage for the same amount of energy deposited in living tissue.

The relative effectiveness of various kinds of radiation in causing biological damage is known as "relative biological effectiveness" (RBE). Over the decades, medical estimates of the dangers of internal alpha exposure have increased with more research. Until the mid-1980s, it was common to use an RBE of 10 for alpha radiation.[21] Since that time, the International Commission on Radiation Protection has recommended that this be increased to 20. (By comparison, gamma radiation has an RBE of 1.)

Very recent research has heightened concern that the true biological damage of alpha radiation may be even higher. Through *in vitro* studies of mouse hematopoietic stem cell colonies, Kadhim et al. found that exposure to a small number of alpha particles (but not X-rays) produced a high frequency of non-clonal aberrations in clonal descendants. This suggests that individual surviving stem cells can transmit to their progeny cells a chromosomal instability that can result in a variety of visible cytogenetic aberrations many cell cycles later.[22] It is well known, in turn, that humans with similar chromoso-

20 Physicians for Social Responsibility 1992.

21 The energy deposited per unit of mass in a medium is measured in units of grays or rads (1 gray = 100 rads), while the biological damage is measured in sieverts or rems (1 sievert = 100 rems). See glossary for fuller definitions of these units.

22 Kadhim et al. 1992.

mal instability defects are more prone to the development of early cancers.[23] This type of transmitted defect is quite distinct from stably induced somatic mutations, which are clonal and readily induced by low LET radiation.

In addition, Nagasawa and Little found that alpha particles at a dose of 0.31 mGy (31 millirads) caused a significant increase in the frequency of sister chromatid exchanges, a marker of genetic damage, in Chinese hamster ovary cells irradiated in the G_1 phase of the cell cycle.[24] A dose of approximately 2.0 Gy was necessary to produce a similar increase in exchanges by X-rays.

These studies suggest that plutonium either has a higher RBE than previously calculated or is more carcinogenic than would be predicted by traditional RBE calculations. If confirmed, this research has implications for both the setting of standards for allowable exposure to plutonium as well as the design and interpretation of epidemiological studies of populations exposed to plutonium.

ROUTES OF EXPOSURE AND BIOKINETICS

In addition to level of dose, the toxicity of plutonium depends on route of exposure, particle size, chemical form, and isotope. The route of exposure of greatest concern is inhalation. Once inhaled, plutonium can become lodged in the sensitive tissues of the lung. Studies in humans and beagle dogs have indicated that such deposits of plutonium remain for years, with gradual absorption into the circulation.[25]

Outside of the body, plutonium is usually less dangerous than gamma-radiation sources. Since alpha particles have a very short range, plutonium on or near the skin deposits essentially all of its energy in the outer, non-living layer of the skin, where it does not cause biological damage. The gamma photons emitted from plutonium decay penetrate the body, but as these are relatively few and weak, a considerable quantity of plutonium would be necessary to yield substantial doses of gamma radiation.[26] (For this reason, plutonium can

---

23 Evans 1992.

24 H. Nagasawa and J.B. Little. 1992. Induction of sister chromatid exchanges by extremely low doses of alpha particles. *Cancer Research* 52: 6394-6396.

25 Voelz et al. 1976; Thompson 1989; Cuddihy et al. 1976.

26 However, gamma radiation from plutonium increases with age due to the presence of small quantities of plutonium-241 (as an unavoidable contaminant). Plutonium-241 (half life 14 years) decays into americium-241 by emitting a beta particle. Since americium-241 has a far longer half-life (432 years), it builds up as plutonium-241 decays. Therefore, the gamma radiation from americium-241 decay, which is far stronger than that from plutonium-239, also builds up with age.

be transported with minimal shielding, with no danger of immediate serious radiological effects.) A wound, however, would render skin more vulnerable. Studies of beagles indicate that a significant amount of plutonium can be absorbed from a skin wound and enter the general circulatory system.[27]

Ingestion of plutonium is a possible route of exposure, through hand-to-mouth transfer of plutonium-contaminated soil or the consumption of contaminated food and water. However, the gastrointestinal absorption of plutonium oxide is less than 0.1 percent,[28] and the greater part of ingested plutonium is rapidly excreted.

Given the same total amount, plutonium is more dangerous in the form of fine particles than as large ones. When large particles (greater than 5–10 microns) are inhaled, they tend to be trapped in nasal hair or deposited on the surfaces of the bronchial airways, where they can be disposed of by the normal clearance mechanisms of the respiratory tree. These particles are then either ingested, which leads to little, if any, absorption, or excreted by coughing or spitting. Smaller particles (less than 1 micron), however, gain entry into alveoli (terminal air sacs of the lung), where they can become lodged, irradiating the surrounding tissue.

Retained plutonium is gradually absorbed, distributed throughout the body, and excreted via urine. Beagle studies have demonstrated that most plutonium retained in the lung is transferred to pulmonary lymph nodes within months to years. Plutonium is also distributed to hepatic and splenic lymph nodes, ovaries, kidney, other soft tissues, bone, and teeth.[29]

Much of plutonium biokinetics (i.e., rates of absorption and excretion, proportion of tissue distribution, etc.) depends on the chemical form of plutonium. Soluble forms of plutonium, e.g., plutonium nitrate, are absorbed from lung relatively rapidly and are deposited heavily in bone and liver, whereas most of the relatively insoluble plutonium oxide is retained in the lung for years, with gradual internal translocation to pulmonary lymph nodes.[30] Half of deposited plutonium oxide is distributed out of the lung by 4 years, with 75 percent of extraplumonary deposits found in the liver and 21 percent in bone.[31]

27 Dagle et al. 1984.

28 Bair 1975.

29 Thompson 1989; Park et al. 1972; Jee and Arnold 1960.

30 Thompson 1989.

31 Cuddihy et al. 1976.

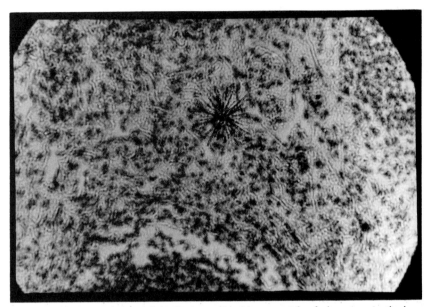

Figure 1.2. Tracks made by alpha radiation emitted by a particle of plutonium in the lung tissue of an ape, magnified 500 times. Photo by Robert Del Tredici.

Unlike radium, another bone-seeking element, which tends to be incorporated exclusively into the calcified mineral matrix of bone, plutonium has an affinity for the non-calcified, non-cartilaginous areas of bone, including the epiphyseum (bone growth plate), the periosteum (outer bone skin), and the endosteum (inner bone in contact with marrow).[32] Deposition is predominantly in trabecular bone (spongy bone in vertebrae and at ends of long bones) rather than in cortical bone.

Species and age are additional factors determining the biological effect of plutonium. For example, younger animals deposit a proportionately larger amount of absorbed plutonium in bone. Studies on monkeys have demonstrated that plutonium deposits in bone concentrate on endosteal surfaces.[33]

Some data are available on the biokinetics of plutonium in humans. In workers who accidentally inhaled plutonium-238 oxide in an insoluble matrix, plutonium was observed to appear in urine within six weeks of

32 Hamilton 1949.
33 Durbin and Jeung 1976.
34 International Commission on Radiological Protection 1972.
35 Voelz et al. 1976.

exposure[34] and then remained measurable in urine for years.[35] Whole body counting cannot be used to estimate body plutonium burden because alpha radiation does not penetrate the skin. Attempts have been made to estimate total plutonium body burden from urinary concentrations and *in vivo* chest counts of plutonium's weak 17-kilovolt X-rays or gamma rays; great variability seems to exist in the relatively sparse data, however, making accurate extrapolation difficult.[36]

The Plutonium Injection Experiment on Humans

Other data on humans derive from an experiment that was begun in April 1945 and carried out on chronically ill patients by Los Alamos National Laboratory in collaboration with the Rochester School of Medicine and Dentistry. The purpose was to gain "adequate information as to the fixation and excretion of plutonium by man [which] is essential to the evaluation and interpretation of the maximum permissible body tolerance."[37] Twelve ill patients were chosen for the experiment whom the authors stated were

> suffering from chronic disorders such that survival for ten years was highly improbable. By adhering to these criteria, the possibility of late radiation effects would be avoided. Furthermore, an opportunity to obtain postmortem material within a few months, or at most a few years, would be much greater.[38]

Two of the subjects were under 45, the youngest being an 18-year-old female with Cushing's syndrome. Each subject was injected with plutonium in the form of plutonium citrate in amounts ranging from 4.6 to 6.5 micrograms. While the subjects were alive, regular physical examinations were performed, and blood, urine and fecal specimens were collected for plutonium measurements and standard clinical assays. At the time of death, samples were collected and analyzed at autopsy.

During the course of the study, the authors did not perceive any sign of clinical toxicity in either the clinical exams or laboratory tests. Monitoring of urine and fecal excretion of plutonium permitted the estimation of the half-life of plutonium in the body — 118 years — as well as the observation that distribution kinetics of plutonium in the human body do not differ substantially from those in animals. The long half-life of plutonium suggests that once absorbed, plutonium poses a lifetime risk due to its neglible excretion rate.

It is difficult, however, to reach firm conclusions from these experi-

36 Voelz et al. 1976.
37 Langham et al. 1950, p. 9.
38 Langham et al. 1950, p. 10.

ments. Inter-individual variations in the observed data were large. The study was performed on ill, mostly elderly subjects who can be expected to have had metabolisms much different than those of young, healthy people. Finally, we must condemn this experiment as unethical. No therapeutic benefits to the patients were intended, and scientists knew of the toxicity of plutonium even then. Informed consent of the patients was not obtained (since even the word "plutonium" was classified during World War II); surviving patients were only told of their injection with plutonium in 1974.

CARCINOGENICITY OF PLUTONIUM: ANIMAL STUDIES

Experiments on beagles have shown that a very small amount of plutonium inhaled in relatively insoluble form, such as plutonium oxide, will with high probability produce lung cancer. In some experiments, lung tumors arose in 100 percent of the animals. These tumors are predominantly bronchioalveolar carcinomas originating in areas of fibrosis and cell abnormalities in peripheral lung where plutonium is deposited. Extrapolating the dose-response data from these experiments to humans, we estimate that the lung dose of plutonium-239 needed to cause lung cancer with a probability approaching 100 percent is about 27 micrograms, or about a millionth of an ounce.[39] Fetter and von Hippel estimated that a single inhalation of 80 micrograms of weapon-grade plutonium (6 percent Pu-240 and 94 percent Pu-239), of which 15 percent would be retained, would lead to a 100 percent risk of death from lung cancer.[40]

Soluble forms of plutonium that have greater systemic absorption, such as plutonium-238 oxide, were found to produce bone tumors in dogs, primarily

---

39 See, for example, Bair and Thompson 1974. This article reports that 0.003 microcuries of plutonium (in the form of plutonium-239 dioxide particles under 10 microns) deposited per gram of lung is enough to cause bronchio-alveolar cancer (a relatively less common form of lung cancer) in 100 percent of cases of exposed beagle dogs. Thus, 0.003 microcuries per gram x 570 grams of lung per human x 16 micrograms of plutonium-239 per microcurie = 27.4 micrograms to cause lung cancer in the average adult human. However, since this experiment was inadvertently a saturation experiment (i.e., all the dogs, including the lowest dose recipient, got lung cancer), the actual risk is probably higher.

   The amount of plutonium needed to cause cancer may be smaller in children. In addition, the risk to smokers may well be much higher, as it is for radon exposure, because of synergy.

   Also see McClellan, particularly the article in it entitled "Status Report: Toxicity of Inhaled Alpha-Emitting Radionuclides," by B.A. Muggenburg et al.

40 S. Fetter and F. von Hippel. 1990. The hazard from plutonium dispersal by nuclear-warhead accidents. *Science and Global Security* 2: 21-41.

osteogenic sarcomas. These tumors originated predominantly in trabecular bone, usually in the long bones or vertebrae.[41] Incident plutonium dose and rate of bone turnover are factors increasing the risk of osteosarcoma in particular bone sites.

CARCINOGENICITY OF PLUTONIUM: HUMAN STUDIES

Few published studies exist from which one can directly estimate the carcinogenic risk of plutonium in humans. Most relevant published studies have been on cohorts of workers involved in nuclear weapons production who were exposed to multiple sources of radiation in addition to plutonium. Other obstacles to using even these studies to estimate plutonium risk are typical of those encountered in environmental epidemiology:

- uncertainties in identifying exposure times and dose based on records, leading to exposure misclassification
- the difficulty of measuring plutonium in the body and the lack of surrogate biological markers of exposure
- inter-individual variation in the metabolism and excretion of plutonium
- inadequate control of potential confounders, such as smoking, in epidemiological studies
- inadequate follow-up of the morbidity and mortality experienced by a population being studied (e.g., loss to follow-up of retired or transferred workers in occupational studies).

Some of these obstacles are difficult to address in any epidemiological investigation; others, such as follow-up investigation of the morbidity and mortality experience of a study population, require a diligence and concern that were likely absent in the nuclear weapons industry.

Regarding this last point, in 1975, 30 years after large amounts of plutonium began to be handled, thereby causing some exposures to the workers who dealt with it, Dr. George Voelz, the medical director of Los Alamos Scientific Laboratory, noted:

> Formal studies for delayed effects from these [plutonium] exposures have not been reported, so it is only possible to state that no cases of acute human pathology following plutonium exposures have been reported to date. Most of these workers have been followed with regular periodic medical examinations during their employment with AEC contractors. *After termination of employment most workers have not been followed by medical examinations for the*

41 Thompson 1989.

specific purpose of determining possible clinical effects from plutonium (or any other hazardous materials they may have encountered in their work). . . .

It would be nice to be able to report that the long-term studies on plutonium workers have been practiced faithfully throughout the industry. *Unfortunately, the follow-up of workers following termination of their employment in plutonium work has been limited to only a few special situations.* [emphasis added][42]

This paucity of available data on the effects of human exposure to plutonium is both unfortunate and inexcusable. It is unfortunate because it forces plutonium risk estimates to rely on animal studies (which are valuable, but extrapolation to humans is always uncertain) and on human studies with small sample sizes (which means that the sensitivity of the study is low and the uncertainty of the results large). And it is inexcusable given the large number of plutonium workers employed over the last five decades in the U.S. alone, on whom data in fact exist. As concluded by the recent review of the Physicians for Social Responsibility task force, the U.S. government and its contractors have simply failed to set up the studies to properly collect and analyze these data.

Finally, it is noteworthy that, despite the fragmentary and flawed nature of the research that has been performed, the PSR task force reviewing studies of nuclear weapons industry workers in the U.S. identified several cancer types for which five or more study populations had demonstrated a standardized mortality or incidence ratio greater than one (and the occurrence of at least five cases), or a standardized ratio that was significantly higher than expected at the $p < 0.1$ level, or a statistically significant increase in cancer with increased radiation exposure.[43] These cancer types included lymphatic and hematopoietic cancers, non-Hodgkins lymphoma, brain and central nervous system cancer, prostate cancer, and lung cancer.

One of the few attempts to fully follow a cohort of workers exposed mainly to plutonium is a long-term study of 26 white males from the Manhattan Project exposed to plutonium at Los Alamos in 1944 and 1945, where the first nuclear weapons were made. Studies of their health status have been periodically published, most recently in 1991.[44]

The amounts of plutonium deposited in the bodies of the subjects were estimated to range from "a low of 110 Bq (3 nCi) . . . up to 6960 Bq (188

42 Voelz 1975.
43 Physicians for Social Responsibility 1992.
44 Voelz and Lawrence 1991.

nCi)."[45] These quantities corresponded to a weight range of 0.043 micrograms to 3 micrograms. Neither the lung dose initially received nor the route of exposure which resulted in the plutonium body burden is known with certainty. The solubility characteristics of the inhaled plutonium are also not well understood, thereby creating uncertainty as to which organs of the body are being irradiated (insoluble particles stay trapped in the lung for a long time, whereas soluble plutonium is relatively quickly metabolized and translocates to other organs such as bone) and at what dose.[46]

Seven of the subjects had died by 1990. The listed causes of death were three lung cancers (including one where the cause of death was listed as heart disease, but the underlying cause was lung cancer), one bone cancer (bone sarcoma), one myocardial infarction, one pneumonia/heart failure, and one auto-pedestrian accident.[47] While four of the seven deaths were due to cancer, little can be inferred from these small numbers. Interpretation is also complicated by the fact that all three people who had lung cancer had smoked cigarettes.

Unlike lung cancer, however, bone cancer is rare in humans. Its expected occurrence in a group of 26 men over a 40-year timeframe is only about one in 100.[48] The plutonium worker's bone cancer occurred in the sacrum and was diagnosed in 1989, allowing a latency period of 43 years after his exposure. Its occurrence among a population of this size (where the subject, incidentally, received a plutonium dose below that of current occupational radiation protection guidelines)[49] is suggestive, especially in view of plutonium's affinity for bone.

Any other inference from this study is very difficult. Obviously, the small sample size severely limits the power of this study to detect anything but the most grotesquely elevated cancer risk. Nevertheless, this study is one of the very few that has attempted full follow-up of an exposed cohort. The

---

45 Voelz and Lawrence 1991, p. 186.

46 These aspects of the study are discussed in some detail in Gofman 1981, pp. 510–520 (based on the status of the Manhattan Project workers study as published in Voelz et al. 1979). Gofman notes evidence indicating that the inhaled plutonium was principally in an aerosolized, dissolved form rather than in insoluble particulates, and he concludes that "nothing in Voelz's entire paper... rules out the possibility that these 26 workers inhaled *only* highly soluble plutonium. If that was the case, the Voelz study is irrelevant to the lung cancer hazard of plutonium *particulates*." (Gofman 1981, p. 516.) We note, however, that it would not be irrelevant to the study of other cancer hazards such as bone sarcomas.

47 Voelz and Lawrence 1991, Table 7.

48 Voelz and Lawrence 1991, p. 189.

49 Voelz and Lawrence 1991.

failure of its authors to comment on the lack of statistical power afforded by a sample size of 26 is unfortunately shared by the authors of most of the other occupational mortality studies of the nuclear weapons industry.[50]

Environmental Regulatory Considerations

The danger to human health posed by small quantities of plutonium has given rise to serious concerns about the various ways in which plutonium contaminates soil, water, and air, and the pathways by which it could reach human beings. These concerns have led to restrictions on plutonium and other transuranic materials in radioactive wastes. Notable among these is the special classification for waste materials containing large quantities of transuranic materials.

The maximum amount of plutonium-239 allowed by U.S. regulations in the air for an off-site population is 2×10^{-5} picocuries per liter. The U.S. Nuclear Regulatory Commission calculates that a person exposed to such a concentration for one year would get an effective dose equivalent of 0.5 millisievert (50 millirems). The corresponding limit for plutonium-239 in water is 20 picocuries per liter.[51] The lower allowable concentration of plutonium in air is due in large part to the higher biological uptake through inhalation than through ingestion; an additional reason is the relatively larger volume of air people breathe each year compared to the volume of water consumed.

The U.S. Environmental Protection Agency also has suggested environmental "action levels" to be used in the clean-up of plutonium-contaminated soil.[52] The EPA's principal suggested action level for newly deposit-

50 Physicians for Social Responsibility 1992.

51 U.S. Nuclear Regulatory Commission 1991, Revised 10 CFR Part 20, Appendix B, Table 2. For a summary of standards for radionuclides in air and water, see Saleska 1992a.

52 An "action level" is a guide used to indicate the need for further study of the situation and for the possible initiation of protective actions and restrictions; it is not an enforceable regulatory limit.

53 This is based on a 1 rem dose to the lung over the course of one year due to resuspension of plutonium particles in air and breathing such air. It is conservative in that it assumes 100 percent occupancy for the full year and a resuspension rate derived from the behavior of relatively newly deposited contamination, which is much more mobile and more easily resuspended than old or stabilized contamination (which EPA says may have resuspension rates as much as 1,000 lower than that used to derive the 0.1 microcurie action level). (U.S. Environmental Protection Agency 1990, Vol. 2, pp. 5–11 and 5–12.)

Table 1.1. Inventory of plutonium contamination in soil for selected sites in the U.S.

LOCATION	APPROX. INVENTORY	REMARKS
Hanford Reservation[a] (central Washington)	6.2×10^{14} Bq (16,700 Ci)	Pu production facility (and other activities)
Nevada Test Site (near Las Vegas, Nevada)	$> 5.7 \times 10^{12}$ Bq ($>$ 155 Ci)	Nuclear test site surface and subsurface tests
Rocky Flats Plant (near Denver, Colorado)	2.9-3.7×10^{11} Bq (8-10 Ci)	Weapons fabrication plant (limited cleanup in progress)
Mound Laboratory (Miamisburg, Ohio)	1.8-2.2×10^{11} Bq (5-6 Ci)[b]	Pu-238 in sediments in canals
Savannah River Plant (southwest South Carolina)	1.1-1.8×10^{11} Bq (3-5 Ci)	Pu and higher isotopes production
Los Alamos Lab (northwest of Santa Fe, New Mexico)	3.7-7.4×10^{10} Bq (1-2 Ci)	Weapons development (high levels in remote canyons)
Trinity Site (near Alamogordo, New Mexico)	1.6×10^{12} Bq (45 Ci)	Site of first atomic bomb test

Source: U.S. Environmental Protection Agency 1990, Vol. 1, Table 1-2, p. 1-11.
Notes:
(a) Total estimated transuranic alpha activity. (U.S. Department of Energy 1991b.)
(b) U.S. Department of Energy (1991b, Table 3.4, p.86) reports a total transuranic alpha activity in soil at Mound of 40 curies.

ed plutonium-239 is 0.1 microcurie per square meter,[53] with a preliminary "screening level" of 0.2 microcurie per square meter in the top centimeter of soil.[54] This matter has assumed some importance because of the contamination of large quantities of soil by plutonium from nuclear weapons production and testing. Areas with plutonium contamination exceeding this level would have to be cleaned up, by removal of topsoil.

Table 1.1 shows the plutonium contamination of various areas in the U.S. nuclear weapons complex, according to official data.

The U.S. Department of Energy classifies wastes containing large quantities of transuranic elements (mainly plutonium, but americium and nep-

54 U.S. Environmental Protection Agency 1990, Vol. 2, p. 3–9.

tunium are important constituents, also) into a special category called "transuranic waste." Such waste must, by law, be disposed of in a deep underground repository. Until the mid-1980s, the definition of transuranic radioactive waste (TRU waste) was waste containing more than 10 nanocuries per gram (370 becquerels per gram) of transuranics. In the mid-1980s the DOE changed the classification of TRU waste to all wastes containing more than 100 nanocuries per gram (3,700 becquerels per gram) of transuranics. This had the effect of reducing the amount of waste that had to be disposed of deep underground.

Summary

Plutonium is a man-made radioactive substance, central to the production of modern nuclear weapons, that poses an extraordinarily dangerous threat to health as an emitter of alpha particles. Experiments in animals have demonstrated that plutonium is readily absorbed when inhaled as fine particles. Absorbed plutonium lingers in the body for decades. Major sites of retention include the lung, lymph nodes, liver, and bone, with relative distribution of the plutonium depending on its chemical form and entry route. Exposed animals develop high rates of cancer, primarily of lung and bone, even when the dose of plutonium is small. Cell culture experiments suggest that such carcinogenesis may reflect a unique ability of alpha particles to cause inherited chromosomal defects from a minute amount of exposure.

The scant amount of data that exist on humans suggests that the behavior of plutonium in the body is similar to that in animals. Well-designed epidemiological studies are lacking due in part to the failure of the nuclear industry to attend to this critical need. Nevertheless, a safe conclusion is that plutonium is probably the most carcinogenic substance known. Exposure to this nuclear poison must be prevented.

In spite of its dangers and in part because of them, plutonium confers almost transcendental power in today's world, as did gold in the period of European expansion. With the consequences of past production still to be fully reckoned and dealt with, plutonium continues to be produced, stockpiled, and transported in massive quantities for both military and ostensibly civilian purposes, confronting the world with grave environmental and security risks.

Chapter 2
Plutonium Production and Use

THE PRINCIPAL ENDS for which plutonium has been sought are the manufacture of nuclear weapons and the generation of electric power. The international nuclear establishment commonly makes a distinction between "military" and "civilian" plutonium, the former being generally higher purity plutonium-239 intended for bombs and the latter generally lower purity plutonium-239 for power production. We consider the distinction to be strained for a number of reasons:

1. In spite of the fact that high-purity plutonium-239 is considered the best kind for making nuclear bombs, other, less pure grades of plutonium will do the job.[1]

2. Plutonium is created by irradiating uranium-238 in a nuclear reactor. Although reactors tend to be designed differently depending on the specific purpose, the important differences between military and civilian reactor technology lie mostly in the *mode of operation* and not in the fundamentals of the technology. Thus, it is generally possible to operate an ostensibly civilian reactor in a manner that will allow the production of high-purity plutonium, for example, by irradiating the fuel for a shorter period. In fact, some reactors are dual purpose reactors — that is, they are operated for the production of both power and plutonium.[2]

3. The plutonium categories "military" and "civilian" mean different things in different countries. Generally, the less pure grades of plutonium can be considered "civilian" (because they are less suitable for bombs) and

---

1 Mark et al. 1987; Rowen 1991.

2 The U.S. had one reactor — the N-reactor at Hanford — which was dual purpose. Several of Britain's Magnox reactors have also been used both for electricity generation and plutonium production. The same is apparently true of some reactors in the former Soviet Union.

the higher purity grades "military". However, in the U.S., for example, the DOE's Defense Programs division has in the past produced both fuel-grade and weapons-grade plutonium, yet most of it would be classed as "military" plutonium since it is produced as part of DOE's military program.[3]

4. Especially in those countries (such as U.K. and the former Soviet Union) where nuclear power production and nuclear weapons materials production are carried out by the same agencies, at the same sites, sometimes in the same reactors and reprocessing plants, the true intended destiny of plutonium seems difficult to ascertain and may change depending on geopolitical circumstances and the programs of new national governments.

Nevertheless, the two categories are in general usage by the nuclear states and the international nuclear agencies, so we have decided to divide our discussion of plutonium stocks into these two categories, as well. However, we use the term "presently-civilian" plutonium in order to call attention to the above caveats and especially to the fact that even plutonium that is not "weapons-grade" can be used to make a functional weapon, albeit a less efficient one than might be manufactured with plutonium of greater purity. Furthermore, even a small amount of plutonium, insufficient for making a *nuclear* weapon, can be used as a *radiological* weapon to terrorize populations (as discussed in Chapter 8).

This chapter addresses principles of plutonium fission and its use in nuclear weapons, how plutonium is made, stocks of military plutonium by country, stocks of presently-civilian plutonium and the prospects for generating electricity from plutonium breeder reactors, and hazards associated with making weapons from plutonium.

Plutonium as Fissile Material

Both uranium-235 and plutonium-239 are fissile materials usable for making nuclear weapons and generating electricity. For a variety of reasons, plutonium-239 is more desirable for producing nuclear weapons than is

---

3   For example, during most of the period from the early 1970s to 1983, the DOE's N-reactor at Hanford operated in a fuel-grade production mode. For many years, the DOE planned to build a plant (the Special Isotope Separation plant) to separate the plutonium isotopes and "enrich" this fuel-grade plutonium to weapons-grade. These plans have been indefinitely deferred.

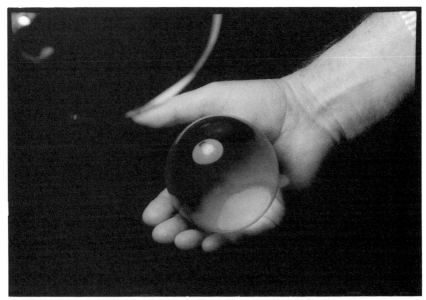

Figure 2.1. This glass ball, about 8 cm. in diameter, is the size of the plutonium core in the bomb exploded over Nagasaki. Photo by Robert Del Tredici.

uranium-235,[4] and most of the nuclear explosives in the world today are believed to have plutonium-239 as their primary nuclear explosive. As will be discussed below in the section on presently-civilian plutonium, however, the role of plutonium in generating power has fallen far short of the extravagant dreams of its proponents, and most nuclear power reactors today are still fueled by uranium-235.

As mentioned previously, the reason plutonium is useful for weapons and power is the fact that it is one of the few materials (along with uranium-235) with the ability to sustain a fission chain reaction that can be produced in significant quantity. In such a reaction, an atom of the fissile fuel material is split by a neutron. Each splitting, or fission, releases energy and at least one new neutron, which can be used to continue the reaction.

In general, the amount of fissile material needed for such a self-sustaining nuclear chain reaction is called a "critical mass." This is not a fixed number but depends on the chemical form of the fissile material (metal or oxide), the shape in which it is arranged, the density to which it is compressed, and the presence of neutron reflectors or moderators (such as

4 For example, it takes a smaller amount of plutonium to achieve criticality under given conditions, so plutonium bombs can be made more compact for a given yield.

water). The form of the material affects its density; pure plutonium metal, which is used for making nuclear weapons, is denser than the oxide form, which is more suited for nuclear power production. Lower density material requires greater mass to achieve criticality. The optimum shape for criticality is spherical; other shapes increase the critical mass (and some shapes prevent criticality altogether).[5] If fissile material is surrounded by a neutron-reflecting material, such as water or beryllium, the mass required for criticality can be reduced by a factor of about two.[6] The amount of plutonium required for bare criticality of an uncompressed plutonium sphere is about 11 kilograms (compared to 50 kilograms of uranium-235 required for criticality), but this can vary widely depending on the factors mentioned.[7] Under optimal conditions of neutron moderation and reflection, the critical mass of a plutonium-239 sphere can be as low as about 500 grams.

The principal difference between a nuclear explosion and a nuclear power plant is the rate of the nuclear chain reaction and the rate at which the energy is released. In a nuclear power plant the energy is released relatively slowly, in a controlled nuclear reaction. The rate of neutron production is carefully regulated so that just a sufficient number of neutrons are available to sustain the chain reaction. With a nuclear weapon, the energy must be released within microseconds in order to create an explosion. In a typical fission bomb, the plutonium is compressed into a supercritical mass (several times that required for bare criticality) using a conventional high explosive. Once the reaction starts, the number of fissions (and resultant neutrons) escalates rapidly, producing a nuclear blast.

The more perfect the implosion and the higher the compression forces created, the smaller the amount of plutonium that can be driven to a critical density. Nuclear weapons designers try to reduce the amount of plutonium needed to produce an explosion of a given strength by achieving very high

---

5 For example, for highly enriched uranium nitrate, criticality cannot be achieved in a cylinder of diameter 6 inches or less no matter how high the concentration or the length of such a cylinder. Geometric dependence of critical mass is used in the design of process vessels in weapons plants to reduce the possibility of accidental criticality. (U.S. National Academy of Sciences 1989, pp. 115–116.)

6 "Neutron moderation" refers to the slowing down of neutrons. Graphite and water are commonly used in nuclear reactors to slow neutrons. As neutrons slow (up to a certain point), the probability that they will be absorbed and cause a fission (in fissile materials such as plutonium-239 or uranium-235) is increased. Thus, neutron moderation decreases the mass of fissile material required to achieve criticality.

7 U.S. National Academy of Sciences 1989, p. 115.

compression of plutonium, and by "boosting," which increases the number of neutrons available to cause fissions by means of the tritium-deuterium reaction.[8] (Boosted nuclear weapons require tritium, which is produced in a manner similar to plutonium (see below).) Typical plutonium-based nuclear weapons contain roughly three to five kilograms of plutonium.

Plutonium Production Technology

There are several important stages in the production of nuclear weapons containing plutonium. The first and major technological hurdle is the plutonium production itself, which is discussed in this section. This is followed by the processing and machining of the plutonium into parts for weapons and the assembly of these parts, which is discussed in a subsequent section.

There are two key facilities needed for the production of plutonium: a nuclear reactor, which produces the man-made element plutonium from the uranium-238 in target fuel rods during the course of the controlled nuclear reaction in its core (reactors used for producing plutonium and tritium are called production reactors), and a chemical separation or reprocessing plant to chemically separate the plutonium from the irradiated reactor fuel after it is removed from the reactor.

NUCLEAR REACTORS AND PLUTONIUM GRADES

All uranium-fueled reactors produce at least some amount of plutonium as a result of the presence of the uranium-238 isotope in the fuel.[9] Some of the uranium-238 in the fuel rods is converted into plutonium-239 as a result of neutron absorption. Gradually, some of the plutonium-239, in

8 Boosting (incorporated into almost all U.S. weapons since 1960) is the principal method of increasing the yield per unit of plutonium used in modern weapons. Boosted weapons contain tritium and deuterium gas, which is released into the pit of the weapon just before firing. As the fission chain reaction proceeds, it heats the tritium-deuterium gas mixture to the point (about 10 million degrees C) where a thermonuclear reaction is initiated. The tritium-deuterium thermonuclear reaction in boosted weapons does not contribute a large amount of energy itself, but because it proceeds so rapidly, it does inject a near instantaneous pulse of neutrons into the plutonium fission reaction. This increases the rate of fissions very sharply, resulting in a several-fold increase in total energy released. (U.S. National Academy of Sciences 1989, p. 126.)

9 The amount of plutonium production is very small, however, in reactors (such as research reactors or naval reactors) fueled by highly-enriched fuel (which consists of almost all uranium-235 and almost no uranium-238).

turn, is converted into plutonium-240 upon the absorption of another neutron. The longer a reactor operates, the more of the uranium-238 is converted to plutonium-239. As the amount of plutonium-239 rises, plutonium-240 begins to build up as well.[10]

(A similar process is used to produce the weapons material tritium, which, as mentioned above, is used to boost the explosive yield of many modern warheads. Tritium results when lithium-6 absorbs a neutron. When tritium production is the purpose of reactor operation, lithium is inserted into the reactor in place of the uranium-238 that would be used to produce plutonium. Nuclear production reactors can generally be used to produce plutonium, tritium, or both.)

Plutonium is produced in both civilian and military nuclear reactors. There are differences in the way the reactors are operated, however, depending on whether the goal is power production or plutonium production.

One key difference is the amount of irradiation or "burn-up" the fuel is subjected to in the reactor before it is removed.[11] In reactors used for power generation, the fuel generally is much more highly irradiated and stays in the reactor much longer than when the purpose is plutonium production for weapons. This is because, as mentioned above, the longer the fuel remains in the reactor, the more contaminated the plutonium-239 becomes with the higher isotopes of plutonium (plutonium-240, -241, -242, etc.) These higher isotopes are generally undesirable from the perspective of the weapons designer because in large amounts they increase the amount of plutonium needed for a bomb and make its yield less predictable. Since the different plutonium isotopes are difficult to separate from each other, the principal means of assuring plutonium purity is to keep fuel irradiation low enough to prevent much build-up of the higher isotopes, especially plutonium-240. When power production is the primary goal, on the other hand, maximizing the energy extracted from each bit of

---

10 This same process of neutron absorption results in the production of some amounts of fissile plutonium-241 from the plutonium-240.

11 Fuel burn-up is a measure of the amount of thermal energy extracted from a given amount of fuel, typically measured in megawatt-days per ton of uranium. A more generic measurement that is sometimes used (in consideration of the fact that the fuel could be plutonium or uranium) is megawatt-days per metric ton of initial heavy metal (MWd/MTIHM). In either case, burn-up is a measure of how much energy was extracted and is therefore related to such factors as the enrichment of the fuel and how long the fuel has been left in the reactor.

12 As reported in Cochran et al. 1987a, p. 136.

fuel (and therefore leaving it in the reactor for as long as practical) is generally more important than the purity of the plutonium in the spent fuel. The plutonium produced in the spent fuel is usually divided into several categories, called "grades," which are distinguished by the purity of the plutonium-239 isotope. The primary distinction is between the purer grades of plutonium (such as weapons grade), and the lower grade categories of fuel grade and reactor grade. The various grades of plutonium are defined by the U.S. DOE according to plutonium-240 content, as follows:[12]

Plutonium Grade	Plutonium-240 Content
Supergrade	2–3%
Weapons grade	<7%
Fuel grade	7–19%
Reactor grade	19% or greater

It is important to remember that, despite the official categories, which imply a sharp distinction between plutonium for use in weapons ("weapons grade") and plutonium for use as fuel in reactors (e.g., "fuel grade"), the categories are somewhat arbitrary. For example, although fuel-grade plutonium is less suitable as bomb material than higher purity mixtures, it still can be used to make a nuclear bomb.[13] This is one of the primary reasons why any consideration of the separation and transportation of plutonium raises significant security concerns, whether or not its ostensible purpose is to support civilian power generation activities.

In sum, lower burn-up means less irradiation and lower production of plutonium-240 and thus higher grade plutonium. Spent fuel in power plants is typically "high burn-up" spent fuel — that is, it is fuel which has been irradiated for extended periods in the reactors so as to generate a large amount of energy. Uranium irradiated for the extraction of plutonium for weapons is "low burn-up" fuel, which has been irradiated for periods at least 10 times shorter than typical power plant fuel. For example, weapons-grade fuel might be irradiated on the order of 1,000 megawatt-days per ton of uranium, while fuel in the most common nuclear power reactors (light-water reactors) might typically be irradiated for 30,000 megawatt-days per ton of uranium.[14]

13 See, for example, Mark et al. 1987 and Rowen 1991, p. 68.

14 This book uses "ton" and "metric ton" interchangeably, always to mean metric ton. A metric ton is 1,000 kilograms, or approximately 2,200 pounds. (A British ton is very close to this, 2,240 pounds. A U.S. ton is 2,000 pounds.)

Another difference between military and civilian reactor operation is that when a reactor is operated for the production of military plutonium, the positioning and amount of uranium-238 is designed to maximize its conversion to plutonium-239. Often, units of natural or depleted uranium are inserted in a production reactor core as *targets*, separate from the uranium-235-containing *fuel*, in order to maximize plutonium production. (Having separate targets and fuel elements also allows for easily switching from plutonium production to tritium production, just by changing the target from uranium-238 to lithium-6.)

Again, it should be emphasized that there is no *fundamental* difference between military and civilian reactor technology.

Producing plutonium in a reactor is only the first step. The plutonium made in a nuclear reactor is mixed up with unconverted uranium and fission products and is unusable in this form. *It is possession of the technology for the next stage of plutonium production — the reprocessing plant — that essentially gives a country the ability to produce plutonium for nuclear weapons.*

REPROCESSING PLANTS

Reprocessing is the chemical separation of plutonium and uranium from fuel which has been irradiated in reactors. It is generally regarded as a key link between civilian nuclear power and nuclear weapons production.

The presence of reprocessing plants is a prime indicator of the ability to make nuclear weapons, whether or not the country in question has a declared program or even the current intention of making them. Thus, all countries that have reprocessing plants should be included in the category of actual or potential nuclear weapons powers.

The specific steps which are needed to recover the plutonium depend somewhat on the type of fuel and reactor, as well as on the choice of processing. The most common chemical separation process is called the Purex process. Purex is an acronym for *Plutonium-URanium EXtraction*. The following steps are involved in the Purex process:

1. *Decladding.* The cladding is removed to expose the contents of the irradiated uranium fuel and/or targets. This may be done chemically, as at Hanford in the U.S. and Sellafield in the U.K., or by a two-step process. In the two-step process, the fuel rods are chopped up into short pieces (shearing), then dissolved (see step 2). A third method applicable to uranium metal fuel rods is mechanical stripping of the cladding from the fuel by longitudinal extrusion through a steel die.

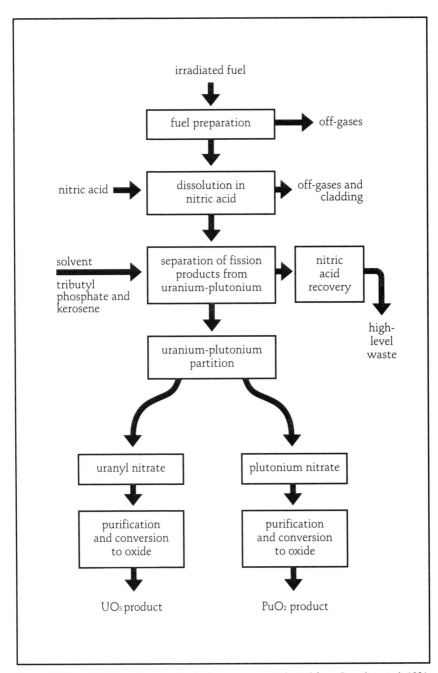

Figure 2.2. Simplified flow diagram for the Purex process. Adapted from Benedict et al. 1981, page 467.

2. *Dissolution of irradiated fuel.* The chopped up fuel rods or the separated fuel rod contents are then dissolved in nitric acid. In the case of chopped up fuel rods, the cladding is separated as part of this process, then collected and stored or discarded as nuclear waste.
3. *Separation of plutonium and uranium from the rest.* The contents of the fuel rods are now in solution as nitrates. This solution is exposed to a solvent called tributyl phosphate (TBP) mixed with kerosene. The TBP selectively separates out the plutonium and uranium from the rest of the contents of the solution.
4. *Separation of plutonium and uranium from each other.* The plutonium and uranium are then separated from each other. The products at this stage are plutonium nitrate and uranium nitrate. These are both in solution.

Plutonium and uranium nitrates may be further processed before shipping to reduce the consequences of transportation accidents. Plutonium in particular is often converted to solid oxide form before shipping. In the U.S., safety regulations now prevent plutonium from being shipped in liquid forms (such as plutonium nitrate solution).[15]

Other processes for the chemical separation of plutonium include the Butex process, which was used at Windscale (now called Sellafield) in England until the 1970s;[16] the Redox (for *RED*uction *OX*idation) process (used at Hanford in the U.S. in the 1950s and 1960s); and the original bismuth phosphate process scaled up from the experimental work done by Glenn Seaborg and his associates as part of the Manhattan Project to build the U.S. atom bomb.

Military Plutonium

In this section, the amounts of plutonium produced for military programs in the countries of the world will be discussed. The subsequent section discusses plutonium production for ostensibly civilian programs. The problems in designating plutonium as "military" or "civilian" that were presented at the beginning of the chapter should be borne firmly in mind throughout.

The world's five declared nuclear weapons states have, of course, produced plutonium for military purposes. These states are the United

15 The U.S. Department of Energy and Nuclear Regulatory Commission instituted a ban on the transport of plutonium in liquid solution on public by-ways effective June 1978.

16 Benedict et al. 1981, p. 461. Also see pp. 459–60 for a history of reprocessing.

States, the former Soviet Union, the United Kingdom, France, and China. In addition, there are the military plutonium inventories of the unofficial but generally recognized nuclear weapons states of India, Pakistan, and Israel. Although Pakistan's nuclear weapons capability is based principally on the use of highly enriched uranium (and not plutonium), it has apparently developed some plutonium production capability as well.[17] We do not have specific data on plutonium inventories of other presumed nuclear or near-nuclear countries such as South Africa or North Korea.

UNITED STATES

The U.S. was the world's first nuclear power, and exploded its first nuclear bomb (using plutonium implosion technology) on July 16, 1945 at Alamogordo, New Mexico. This was followed in August by the enriched uranium bomb dropped on Hiroshima and the plutonium bomb dropped on Nagasaki. The plutonium for these devices was produced for the Manhattan Project by the production reactors at the 365,000-acre Hanford Engineer Works in Washington state (code-named "Site W" during the war). The enriched uranium was produced at the Manhattan Project's other large production complex in Oak Ridge, Tennessee ("Site X"). The bombs were designed and assembled at Los Alamos, New Mexico ("Site Y"). This program was run during the war by the U.S. Army.[18]

In recent decades, the production of nuclear materials and the fabrication of these materials into weapons has been the responsibility of the Department of Energy (DOE), formerly the Atomic Energy Commission (AEC). The DOE now operates about 20 major sites around the U.S., but nuclear production reactors and reprocessing facilities for nuclear materials separation have operated at only two of these: the original Hanford site, now called the Hanford Nuclear Reservation, and the Savannah River site near Aiken, South Carolina. (The Idaho National Engineering Laboratory (INEL) near Idaho Falls, Idaho also reprocesses highly-enriched irradiated fuels but to extract uranium).

Since Hanford's origin, a total of nine production reactors (called B, D, F, H, DR, C, K-East, K-West, and N) and five reprocessing plants (T, B, U, Redox, and Purex) have operated there. The last of these to operate are the

---

17 Pakistan's facility at New Labs is estimated to be capable of separating 10–20 kilograms of plutonium per year. (Spector and Smith 1990, p. 114.) Pakistan has also been reported to be building an unsafeguarded research reactor. (Hibbs 1988.)

18 Sanger and Mull 1989, p. 1.

Figure 2.3. Hanford N-reactor, Richland, Washington, USA, 1981. This graphite-moderated reactor produced plutonium for U.S. nuclear weapons and electricity for commercial use, and is the only such "dual-purpose" reactor in the U.S. It has been closed since 1988. Photo by U.S. Department of Energy.

N-reactor and the Purex reprocessing plant.[19] The N-reactor, a 4,000 megawatt (thermal) (MWt) graphite-moderated reactor is a "dual-purpose" reactor in that it can be used to produce both plutonium and electricity (860 megawatts' worth) for commercial sale at the same time (this is the only reactor operated as a dual-purpose one in the U.S., although there are others in other countries — Britain, for example).[20] The N-reactor was placed on cold stand-by in early 1988 and is not expected to operate again. The Purex plant has not operated since March 1990, and its future is still uncertain;[21] in any case, Hanford's days as a major nuclear materials production center are over.

19 Cochran et al. 1987b, p. 13.

20 It should be noted that the U.S. DOE is considering the construction of a new production reactor, one of the designs for which is dual-purpose.

21 In any case, the only feed for the Hanford Purex plant is a remaining 2,100 tons of N-reactor fuel whose fate has not yet been decided. Even if this fuel is processed, there will be no more produced. Saleska and Makhijani 1990 examine options for managing this remaining fuel.

The other major nuclear production site in the U.S., the Savannah River site, was chosen by the AEC in 1950, constructed and operated by DuPont for most of its history, and is now operated for the DOE by Westinghouse. There are five production reactors at Savannah River (R, P, K, L, and C). R was shut down in 1964, and all have been shut down since the summer of 1988. The DOE has been trying to restart the K reactor for tritium production (there are no immediate plans to produce more military plutonium in the U.S.). Two reprocessing plants (the F and H canyons) have processed irradiated fuel and target elements from the production reactors and from offsite foreign and domestic research reactors.[22]

There are numerous test reactors at the Idaho National Engineering Laboratory (INEL), although these are typically not used for large-scale nuclear materials production purposes but for research.[23] One of the principal missions of INEL is the reprocessing of highly enriched irradiated naval fuels (after they have been used in nuclear-powered submarines and ships). This was done at the Idaho Chemical Processing Plant (located at the INEL site). The extracted highly-enriched uranium was then used to make fuel drivers in the Savannah River reactors.

All reprocessing has been shut down in the U.S. since 1988, a situation that in July 1992 was codified as policy by the U.S. administration.

SOVIET UNION [24]

The military and civilian nuclear fuel cycles are the responsibility of the Russian Ministry for Atomic Energy, generally referred to in Russia as "Minatom." This Ministry is the Russian successor of the USSR Ministry of Atomic Power and Industry (MAPI) (which, until 1989, had the obscure title of the Ministry of Medium Machine Building). Plutonium and tritium for nuclear weapons are produced at three sites: the Mayak Chemical Combine at Chelyabinsk-65 (formerly "Chelyabinsk-40," and sometimes referred to in the West as the Kyshtym complex after a nearby town), on a 230-square-kilometer area about 100 kilometers northwest of the regional capital city of Chelyabinsk in the southern Urals; the Siberian Atomic Power Station in the town of Tomsk-7 (also, "Sev-

---

22 Cochran et al. 1987b, pp. 97, 116.

23 Some plutonium is produced at INEL as part of the research and experimentation on breeder reactors.

24 Unless otherwise noted, based on Cochran and Norris 1992.

ersk"), northwest of Tomsk; and Krasnoyarsk-26, a site near the town of
Dodonovo on the Yenisey River, about 50 kilometers northeast of the
city of Krasnoyarsk.

A total of 14 production reactors were operating at these sites before
1987. Eight of these were shut down by the beginning of 1991. Chelya-
binsk-65 houses one heavy-water-moderated reactor and five graphite-
moderated reactors, the last of which was shut down for decommission-
ing in November 1990. The Siberian Station at Tomsk-7 has five graphite-
moderated reactors (two of which have been shut down since the end of
1990) used for the dual purpose of generating electricity and producing
plutonium for weapons. Krasnoyarsk-26 has the remaining three reactors,
all graphite-moderated reactors for plutonium production only (not tritium). Unlike most sites around the world, all plutonium production
activities at Krasnoyarsk-26 are located deep underground (200 to 250
meters down), apparently to provide protection against potential enemy
attack. Two of the three reactors at Krasnoyarsk-26 are slated to be shut
down by September of 1992, but the third, apparently needed for electricity, is to keep operating until at least the year 2000.

Reactors operating as of May 1992 were the three graphite-moderated
reactors at Krasnoyarsk-26 and three graphite-moderated reactors at
Tomsk-7. The Soviets announced in October 1989 plans to shut down 13
of their plutonium-producing reactors by the year 2000, apparently leaving one reactor for the production of tritium.[25]

The construction of Chelyabinsk-65, the first Soviet plutonium production facility, began along the southeast shore of Lake Kyzyltash in the
upper Techa River drainage basin in November 1945. The first (graphite-
moderated) production reactor started operation there in June 1948.

A chemical reprocessing plant has been operating at Chelyabinsk-65
since December 1948. Until 1978, the plant separated plutonium from
the military production reactors, but since then the irradiated fuel elements from these reactors have been sent to Tomsk for reprocessing.
The Chelyabinsk-65 plant was converted to reprocessing spent fuel from
civilian power reactors and naval reactors to recover plutonium for the
civilian fast breeder reactor program. Tomsk-7 has, in addition to a
chemical reprocessing plant, a uranium enrichment plant. Krasnoyarsk-
26 has a partially finished reprocessing plant, completion of which is
uncertain.

25 Petrovsky 1989.

From assumptions about production activities, an estimate of the Soviet inventory of weapons-grade plutonium is 115–140 tons.[26] This is roughly consistent with an estimate based on krypton-85 discharged to the atmosphere in chemical reprocessing, which indicates that the Soviet plutonium inventory as of 1983 could be as much as 140 tons (with an uncertainty of 25 tons), but since not all reprocessing activities are directed towards weapons grade plutonium production, this is a maximum estimate.[27] Russia is currently producing "civilian" plutonium at a rate of about 2.5 tons per year,[28] with production planned to continue for at least several more years.[29]

UNITED KINGDOM[30]

Britain's main reprocessing center is the Sellafield site (formerly called Windscale)[31] on the Calder River and the Irish Sea. The reprocessing plant is located on the north side of the river, and a series of dual-purpose plutonium-and-electricity-producing reactors are located on the south side. These belong to the first generation of British reactors, known as Magnox reactors, which were fueled by magnesium-alloy-clad natural uranium fuel. (Britain has a second generation of advanced gas-cooled reactors, or AGRs.)

Sellafield reprocesses fuel from the U.K.'s 26 gas-cooled Magnox reactors, 18 of which are at nine sites ostensibly dedicated to "civilian" power generation purposes.[32] Aside from the civilian reactors, there are four plutonium-producing Magnox reactors at Calder Hall, and four similar ones dedicated to plutonium and tritium production at Chapelcross on the west coast of Scotland.[33] Reprocessing began at Sellafield (then Windscale) in 1952 at a plant called B204, which had a capacity of 300 tons of irradiated Magnox fuel per year. This produced the plutonium used in Britain's first nuclear

---

26 Cochran and Norris 1992, pp. 60–61. As explained earlier, we use "ton" to mean "metric ton."

27 von Hippel et al. 1986, Chapter V, p. 6.

28 von Hippel 1992.

29 Cochran and Norris 1992, p. 61.

30 Carter 1987, pp. 238–239; Peden 1991, pp. 69–70.

31 The name was changed in 1981, perhaps in an attempt to make a clean start after a history of mishaps, including a notorious fire in 1957, which released 20,000 curies of radioactive iodine to the atmosphere. (Carter 1987.)

32 Sellafield also reprocesses fuel from a Magnox reactor at Latina in Italy and another Magnox reactor in Japan.

33 Peden 1991.

bomb, exploded in late 1952. In 1964, B204 was shut down, and a new plant, B205 (with a capacity of 1,500 tons per year), began operation. B204 was modified to reprocess the higher burn-up oxide fuels being used in the second-generation AGR reactors, and re-opened in 1969. However, the modified B204 encountered numerous problems and only operated until 1973, when it was permanently closed. In that year, according to an official report by the British government, an unanticipated chemical reaction forced ruthenium gases along a rotor drive shaft which passed through a wall in a hot cell into a working area. From there, ruthenium was then circulated throughout the plant by the ventilation system, causing excess exposures to 34 workers.[34]

This left Sellafield with the capacity to reprocess only Magnox fuel, which is the current state of affairs. However, construction for a new plant was authorized in 1978, and this plant, the Thermal Oxide Reprocessing Plant (THORP), has been operating on a test scale but, due to various delays, may not be fully commissioned until late 1993. This is intended to allow Britain to reprocess oxide fuel from the new generation of AGRs and from foreign light-water reactors.

The U.K. also has a facility at Dounreay used to reprocess fuels from the U.K.'s experimental fast reactor program (see page 44).

FRANCE[35]

France exploded its first nuclear bomb on February 13, 1960. The primary facility for the production of bomb materials in France is the nuclear complex at Marcoule, which began operations in 1955.

The first reprocessing plant was the UP-1 plant at Marcoule, which went into operation in late 1958 and reprocessed irradiated fuels from the early military production reactors at Marcoule. These dual-purpose reactors (designated G1, G2, and G3) were fueled by uranium in metal form, and were of the gas-cooled, graphite-moderated design.[36] Reprocessing facilities at Marcoule currently have a capacity of 800 tons of irradiated fuel per year.

A second reprocessing plant, UP-2, went into operation in 1966 at La Hague, a second nuclear site now devoted principally to the extensive French civilian nuclear industry.[37] UP-2 reprocessed metal fuels (at a capac-

34 Cited in Carter 1987, p. 239.
35 Barrillot 1991.
36 Barrillot 1991; Carter 1987, p. 308.
37 Carter 1987, p. 309.

ity of 800 tons per year) until 1986. In 1976, UP-2 was upgraded to be able to process up to 400 tons per year of the higher burn-up oxide fuels. A third reprocessor, UP-3 (capacity 800 tons), went online in 1989 to fill anticipated reprocessing needs under contracts with Japan and other countries for reprocessing fuel from their electricity-generating plants. In addition, France anticipates increasing the capacity of UP-2 from 400 to 800 tons per year. Thus, La Hague will have a nominal reprocessing capacity of 1,600 tons of irradiated fuel per year, sufficient to separate 10 to 12 tons of plutonium.[38]

CHINA[39]

China tested its first nuclear weapon on October 16, 1964 — four years after the 1960 break which ended Sino-Soviet nuclear cooperation. The bomb was of an implosion design that used enriched uranium (the Chinese did not produce plutonium until later) and had a 20-kiloton yield. After the first test, China moved quickly into thermonuclear designs, and only 32 months later, on June 17, 1967, tested its first hydrogen bomb.

The two production reactors at Lanzhou and Jiuquan (Subei County) are China's main sources of plutonium. The Soviet departure in 1960 brought the Chinese plutonium effort to a temporary halt, and the production reactor at Jiuquan was not completed until 1967. Completion of the reprocessing plant followed in 1970. The Jiuquan Atomic Energy Complex (referred to as Plant 404) also includes the Plutonium Processing Plant for refining plutonium metal, the Nuclear Fuel Processing Plant (which converts uranium hexafluoride to metal), the Nuclear Component Manufacturing Plant (which fabricates nuclear weapon components), and the Assembly Workshop, which does final assembly of weapons.

The decision taken in 1964 to build a duplicate ("third line") set of production facilities in the interior led to the construction of the major production and chemical processing complex at Guangyuan (Sichuan). Also referred to as Plant 821, the Guangyuan complex has China's largest plutonium production reactor and chemical reprocessing plant. It is similar to but larger and newer than the Jiuquan complex.

38 Carter 1987, p. 310.
39 Fieldhouse 1991. See also Lewis and Xue 1988.

ISRAEL[40]

Israel is reported to have a stockpile ranging from some 50 to 200 strategic and tactical nuclear weapons,[41] and perhaps as many as 300.[42] The country's center for the production of nuclear warhead materials is located at Dimona in the Negev desert. The Dimona heavy-water-moderated production reactor and chemical processing plant were supplied by France and built with French assistance, following an agreement concluded in late 1957. The two countries had extremely close ties in the mutual development of their nuclear weapon programs. Construction of the Dimona reactor was well underway in 1960 and operation of the natural uranium-fueled reactor began in 1963. Construction of the plutonium processing plant underground at Dimona was delayed but finally got underway in 1962, and startup occurred in 1966, with full operation by 1972.

The reported thermal power of the Dimona ranges from 40 to 150 megawatts (thermal) (MWt), with the possibility of several upgrades. The power announced by Israel in 1960 was 24 MWt, but, according to the French construction engineers, the cooling capacity was sufficient for a higher power of about 40 MWt. In 1986, Mordecai Vanunu, a technician at the Dimona plant since 1977, provided photographs and detailed information about the operation of the Dimona facility to the London *Sunday Times*. Vanunu claimed an Israeli stockpile of 200 warheads, as opposed to some U.S. government estimates of about 50.[43] Vanunu's detailed account to the *Sunday Times* indicated that over 40 kilograms of plutonium per year were separated in the chemical reprocessing plant. At this rate of plutonium production, about ten nuclear bombs could be produced per year or about 200 bombs in a period of 20 years.

INDIA

In 1974, India detonated a nuclear explosive made from plutonium produced in the 40-MWt Cirus research reactor at the Bhabha Atomic Research Center (BARC) in Trombay. The reactor, which began operation in 1963 and has the capacity to produce some 25 kilograms of plutonium per year, was supplied by Canada. The agreement with Canada restricts

40 Hersh 1991; Spector and Smith 1990.
41 Hersh 1991; Spector and Smith 1990.
42 See Brinkley 1991.
43 Vanunu was abducted by Israeli agents and returned to Israel where he was sentenced in March 1988 to 18 years in a maximum-security prison.

the use of the reactor to peaceful purposes. India did declare the explosion to be for peaceful purposes, and it has claimed not to be building nuclear weapons ever since. However, then-Prime Minister Rajiv Gandhi stated in 1985 that the country could become a nuclear power in a matter of weeks or months, so it is likely that the country has an ongoing nuclear weapons research and development program.[44]

There are indications that India is pursuing a thermonuclear weapons program.[45] Its plutonium production capability received a boost when the 100-MWt Dhruva research reactor became fully operational in 1988. This reactor, which like Cirus is heavy-water-moderated and natural-uranium-fueled, could produce some 25 kilograms of plutonium annually, enough for several weapons a year. The 42-MWt plutonium-fueled fast breeder test reactor, which began operation in 1985, is loaded with a core of 50 kilograms of plutonium.[46] The fast breeder test reactor could produce some 5 kilograms of high-purity (3 percent Pu-240) plutonium annually.[47] Another 52 kilograms of plutonium is in the core of the shutdown Purnima research reactor.[48]

India has two operating fuel reprocessing plants: one at BARC in Trombay, near Bombay, with a capacity of more than 30 tons per year, that processes spent fuel from the Cirus and Dhruva research reactors, and another (Prefre) at Tarapur, with a capacity of 100 tons per year, that processes discharged power reactor fuel. By early 1986, Prefre had started reprocessing spent fuel from the reactors at the Madras power station. A third plant, which is nearing completion at Kalpakkam near Madras, will be able to reprocess 125 tons of spent power reactor fuel annually from the Madras I and II reactors and plutonium fuel from the fast breeder test reactor.

PAKISTAN

Pakistan is believed to have achieved nuclear weapons capability, probably by 1988 if not by 1986.[49] Although Pakistan had apparently actively pursued the plutonium-bomb route, its plans were stalled in 1977 when

44 Albright and Zamora 1989, p. 20.
45 Albright and Zamora 1989, pp. 24, 25.
46 Spector 1985, pp. 281–282.
47 Albright 1988, p. 41.
48 Albright 1988.
49 Spector 1988, p. 120.

Table 2.1. Military plutonium separated as of 1990 (in metric tons)

COUNTRY	FACILITY	QUANTITY
United States	Hanford	60.5 [a]
	Savannah River	47.8 [b]
Soviet Union [c]	Chelyabinsk-65	40.5
	Tomsk-7	53.3
	Krasnoyarsk-26	28.7
United Kingdom [d]	Sellafield	~5.0
France [e]	Marcoule	~6.0
China [f]	Jiuquan (Subei County) and Guangyuan (Sichuan)	1.25 - 2.5
India [g]	Trombay	0.28
Israel [h]	Dimona	0.4 - 0.7
Pakistan [i]	New Labs	¿

Notes:

(a) Includes both weapon- and fuel-grade plutonium produced in the N-reactor before operations halted in 1986. See Cochran et al. 1987a, pp. 64, 65, 75, 76. Uncertainty is +/- 5%.

(b) Cochran et al. 1987a, pp. 63, 75. Uncertainty is +/- 10%.

(c) Cochran and Norris 1992, pp. 59-62. The cumulative figure of 122.5 tons is in plutonium-equivalent, and should be reduced depending on the amount of tritium production. Cochran and Norris 1992 estimate that about 6 tons of plutonium-equivalent has been devoted to producing tritium, leaving an estimated plutonium inventory of about 116 tons. The aggregate figure has an uncertainty of about 20 tons.

(d) Includes 0.78 tons of plutonium produced at Calder Hall and Chapelcross conjectured to have been sent to the U.S. under the Mutual Defense Agreement of 1958. See Hesketh 1984, pp. 42, 85.

(e) Based on Barrillot 1991 and our independent estimates of quantities of French plutonium in weapons and in reserve.

(f) Fieldhouse 1991 and others estimate China's deployed stockpile at 250 to 350 weapons. A rough estimate of the plutonium inventory can be derived from the estimated number of warheads, and by using U.S. averages of quantity of plutonium (5 kg.) per warhead. We take the total number of warheads to be uncertain by a factor of two and to range from 250 to 500. Thus, China's military plutonium inventory is estimated to range from 1,250 kg. to 2,500 kg.

(g) Albright 1988, pp. 43-45.

(h) Based on estimates from information regarding the power of the Dimona reactor revealed by Mordecai Vanunu (see text).

(i) New Labs is capable of separating 10-20 kilograms of plutonium per year. (Spector 1990, p. 114.) Pakistan is reported to be building an unsafeguarded research reactor. (Hibbs 1988.)

France suspended deliveries for a large-scale reprocessing plant Pakistan had contracted to buy. Pakistan thus fell back on highly enriched uranium for nuclear weapon production, which apparently was successfully implemented using gas centrifuge technology built at its Kahuta plant (and apparently obtained with cooperation from officials of the European nuclear industry).[50]

Pakistan now has some plutonium production capability as well, however. It has a small pilot-scale plutonium reprocessing plant called "New Labs" in Rawalpindi. New Labs is believed capable of separating 10–20 kilograms of plutonium per year,[51] but its operational status and the amount of plutonium that might have been produced are not known.

SUMMARY OF MILITARY PLUTONIUM PRODUCTION, BY COUNTRY

Table 2.1 lists the estimated cumulative inventory of plutonium produced for military programs in the eight nuclear-weapons states cited above, along with the names of the facilities that produced it.

Presently-Civilian Plutonium

Plutonium's role in the civilian sector is as a possible fuel for nuclear power reactors. Plutonium fuel for power reactors is often discussed in the context of the proposed use of fast breeder reactors (see below). Plutonium can also be used in existing conventional nuclear reactors in the form of mixed oxide fuel (MOx) in which plutonium is mixed with uranium (see Chapter 7 for discussion). Neither means of using plutonium to fuel electricity generation is in common use due to technical difficulties and cost.

In addition to the declared or undeclared nuclear weapons states discussed in the previous section, several other countries have plutonium separation capability, including Belgium, Japan, and potentially Brazil and Argentina. The latter two have recently stopped their efforts to establish reprocessing capability, however. Commercial reprocessing capacity is listed in Table 2.2.

There is also a commercial trade in reprocessing services for nuclear power, so still other countries own separated plutonium that has been processed elsewhere. For example, many European countries as well as

50 Spector 1988, pp. 135–137.
51 Spector 1988, p. 151.

Table 2.2. Civilian reprocessing capacity

COUNTRY	REPROCESSING PLANT	FUEL TYPE	OPERATIONAL DATES	CAPACITY (metric tons of heavy uranium/yr)
EUROPE & JAPAN				
United Kingdom	B205 (Sellafield)	Metal	1964	1000
France	MAR400+UP1 (Marcoule)	Metal	1958	500
France	UP2 (La Hague)	Metal	1967-1986	800
France	UP2 (LaHague)	Oxide	1976	400
France	UP3 (La Hague)	Oxide	1989	800
France	UP2-800 (La Hague)	Oxide	1993	800
United Kingdom	THORP (Sellafield)	Oxide	1993	700
Japan	Tokai Mura	Oxide	1977	90
Japan	Rokkashomura	Oxide	1999	800
Russia	Chelyabinsk			¿ (a)
Belgium	Eurochemic (Mol)	Oxide	1970-1974	
OTHER COUNTRIES				
India	Prefre (Tarapur)	Oxide	1979	100
India	Kalpakkam	Oxide	1994¿	125
Pakistan	Chasma	Oxide	Suspended¿[b]	100
Argentina	Ezeiza	Oxide	Suspended 1990	5
Brazil	Resende	Oxide	Postponed	2

Sources: Berkhout and Walker 1990, p. 16; Albright 1987; Spector and Smith 1990.
Notes:
(a) Plutonium is being separated at Chelyabinsk-65 at a rate of 2.5 tons per year. (von Hippel 1992.) We do not have figures on capacity in terms of tons of uranium.
(b) France terminated work on the French-supplied plant in 1978, but construction may be continuing. (Spector and Smith 1990, p. 115.)

Japan have sent their spent fuel to France for reprocessing. Table 2.3 lists cumulative separated plutonium by country of ownership and country of production (separation).

BREEDER REACTORS

The dream of nuclear power proponents since the beginning of the nuclear age has been to convert non-fissile, relatively abundant uranium-238 into fissile plutonium in reactors in such a way that the amount of fissile material produced would be larger than the amount of fissile material consumed during the reactor operation. Such a reactor is called a "breeder reactor." Breeder reactors designed to breed plutonium must rely on fast neutrons for maintaining the fission chain reactions, hence the name "fast breeder

Table 2.3. Civilian plutonium separated as of 1990 (in metric tons)

REPROCESSOR COUNTRY:	FRANCE		UK	BELGIUM	GERMANY	USSR	JAPAN	INDIA	US
REPROCESSOR:	La Hague	Marcoule	Sellafield	Mol	Karlsruhe	Chelyabinsk	Tokai	Prefre	West Valley
FUEL TYPE:	oxide	metal	metal	oxide	oxide	oxide	oxide	oxide	oxide
Belgium	1.1			0.68					
France	3.8	12.0							
Germany	14.2				0.94				
India								0.5	
Italy			2.8						
Japan[a]	1.2[b]		1.9[c]				2.9		
Netherlands	0.6								
USSR						25.0			
Spain		4.9							
Switzerland	1.1								
U.K.			42.9[d]						
U.S.									1.33
TOTALS:	22.0	16.9	47.9	0.68	0.94	25.0	2.9	0.5	1.33

(Left margin label: OWNER COUNTRY)

Sources: Berkhout et al. 1990; Berkhout and Walker 1991a, 1991b; Albright 1987, 1988.

Notes:

(a) Does not include an additional 0.640 ton of plutonium purchased and shipped to Japan from the U.S., U.K., France, and Germany.

(b) Includes 0.253 ton of plutonium shipped by sea to Japan in 1984. See Nuke Info Tokyo 1990, p. 2.

(c) Plutonium recovered from Japanese Magnox fuel: 0.92 ton returned to Japan, 0.95 ton stored in U.K.

(d) Includes plutonium produced in U.K. civilian power reactors and non-weapon-grade plutonium produced in Calder Hill and Chapelcross military reactors when optimized for electricity production after 1964. Also includes some 4 tons of plutonium produced in civilian reactors and sent to U.S. in barter after 1964 under the Mutual Defense Agreement of 1958. In addition, includes 300 kg. of plutonium recovered from oxide light-water-reactor fuel. Weapon-grade plutonium produced in the military production reactors is not included.

At Sellafield through 1985, an estimated 44.3 tons of plutonium was separated from all sources, as suggested by Barnham and based on 70 percent of the plutonium in Sellafield waste through that time arising from the recovery of 31 tons of civil-origin Magnox plutonium. (See Keith Barnham's article in Barnaby 1992.)

reactors" is often used for plutonium breeder reactors. In contrast, light-water reactors as well as graphite-moderated reactors use far slower neutrons (called thermal neutrons) to keep the chain reaction going.[52]

Initially there was great optimism about breeder reactors and the possibility of relying on plutonium energy in the civilian sector. However, breeder reactor programs have largely failed, due to both economic shortcomings (which are partly a result of the cheaper than expected price of uranium) and unanticipated technical difficulties in commercial implementation.[53]

Once widespread, programs to commercialize breeder technology have experienced great difficulties, after a total investment of tens of billions of dollars.[54] Currently, the only commercial-size breeder reactor outside of Russia — France's Superphenix — has encountered so many technical problems that in July 1992 the French government took the decision not to permit restart without a further public inquiry. The U.K. invested billions of dollars in its fast breeder reactor program between 1954 and 1988 but only has a prototype fast breeder reactor at Dounreay. Government funding for the prototype breeder reactor will end in 1994, and government funding for reprocessing at Dounreay will end in 1997. Germany's breeder reactor, at Kalkar, was finished in 1986 but failed to get an operating license and was later abandoned. In the U.S., the demonstration breeder reactor that was proposed at Clinch River, Tennessee was canceled in the late 1980s. The once-ambitious Soviet breeder program is now

52 The terms "thermal" and "fast" refer to the average speed of the neutrons in the reactor core. A thermal reactor is one in which the neutrons are slowed to the point where they are in rough thermal equilibrium with the reactor materials (set speeds around 3,000 m./sec.). In a fast reactor, the neutrons are not slowed, and they have speeds close to that which they have at the moment of fission (about 15 million m./sec.). Neutrons are much more effectively absorbed at thermal speeds in fissile materials such as plutonium-239 and uranium-235. For this reason, thermal reactors are the preferred type except when breeding plutonium is the objective. In breeding plutonium, there must be greater than 2 neutrons produced for every neutron absorbed or breeding will not take place (one neutron is required to sustain the reaction, and one to produce an atom of plutonium-239 to replace that which was fissioned). Because fissions induced in uranium-235 and plutonium-239 by thermal neutrons produce less than 2 neutrons per neutron absorbed, fast neutrons must be used. (This is not the case when uranium-233 is used as a fuel to breed more uranium-233 from thorium-232, because uranium-233 produces more than 2 neutrons per thermal neutron absorbed.) (Benedict et al. 1981, pp. 6–7.)

53 The economics of breeder reactors and plutonium recycle are discussed in Albright and Feiveson 1988 and in Berkhout and Walker 1990.

54 Commissariat à l'Energie Atomique 1988; International Energy Agency 1989, p. 94.

Figure 2.4. The Superphenix breeder reactor, Creys-Malville, France, 1981. The Superphenix has been shut down due to technical difficulties. In July 1992 the French government decided not to permit it to start up again until a public inquiry is carried out. Photo by Robert Del Tredici.

stalled, plagued by economic and safety concerns. A 600-megawatt breeder reactor at Beloyarsk, Russia is still running on enriched uranium fuel after 12 years of operation.[55] Progress on two liquid-metal-cooled fast breeder reactors under construction at Chelyabinsk-65 and a planned third has been stopped since 1987. As of mid-1992, separated "civilian" plutonium at Chelyabinsk had accumulated to a reported total of 30 tons and, as mentioned above, is still being produced at a rate of 2.5 tons per year.[56] Even Japan, the only country still counting on breeder reactors, is looking at ways to convert its prototype breeder reactor at Monju, due on line in 1993, to burn rather than breed plutonium.

The fall from favor of breeder technology and plutonium-fuel-cycle nuclear programs may mean that a large inventory of unused and unwanted plutonium is left as a difficult legacy in the civilian sector as well as the military one — an ironic development for a material once viewed as more valuable than gold. The example of Japan is instructive.

55 Personal communication from John Large, of Large and Associates, London, to Katherine Yih, July 1992. Large learned this in Ekaterinsburg (formerly Sverdlovsk) in July 1992.

. 56 von Hippel 1992.

JAPAN'S NUCLEAR POWER PROGRAM[57]

By early in the next century, if current plans are carried out, about half of Japan's electricity is expected to be supplied by nuclear power.[58]

Japan has had a long-term vision of an indigenous energy supply based on breeder reactors and plutonium fuel. This vision is increasingly troubled, however, by international concerns over nuclear proliferation and by economic factors which make breeders and plutonium "recycle" unattractive. In the move toward a plutonium economy, reactor fuel reprocessing is at the center of the long-range nuclear energy plans. Japan now operates a 90-ton-per-year reprocessing plant at Tokai Mura and is building a 800-ton-per-year plant at Rokkashomura for separating plutonium from conventional light-water reactor fuel. In addition, Japan has reprocessing commitments from both Britain and France that extend beyond the year 2000 that would make available some 40 tons of plutonium by the year 2010. If the present trend continues, Japan could become the world's largest user of plutonium, surpassing even France.

Yet the plutonium will not be used as fast as it is separated without massive use in conventional light-water reactors. Japan's commercial breeder program is not expected until 2030 at the earliest and very likely will be delayed further. A plutonium surplus is likely to build up.

By the year 2010, besides the 40 tons of Japanese plutonium that will have been separated at British and French reprocessing plants, another 70 tons will be separated at the Rokkashomura plant, if it operates at full capacity starting in 2000. Out of the 110 tons accumulated, officials claim about 40 will be needed for breeder reactors, and another 13 tons might be used in heavy-water-moderated advanced thermal reactors.[59]

Current plans call for a program of unprecedented magnitude to use the remaining 57 tons as MOx fuel in light-water reactors. The Japan Atomic Energy Commission is expected to make a formal recommendation that most of the recovered plutonium be used in light-water reactors until breeder reactors come into operation.[60] Should the program of full-scale

---

57 Berkhout et al. 1990, pp. 523–543; Suzuki 1991.

58 Swinbanks 1991b.

59 See Suzuki 1991. Atomic Energy Agency officials estimate an inventory of 84 tons of plutonium by the year 2010. (Swinbanks 1991a.) Assuming that 84 tons refers to the fissile content, the total quantity of plutonium is 112 tons, assuming a fissile content of 75 percent.

60 Swinbanks 1991a.

MOx burning falter, the alternatives are to cut back on reprocessing or else face the possibility of a large plutonium surplus.

One of the problems that Japan faces is shipping the plutonium from Europe back to Japan. All avenues but sea transport have been effectively blocked by U.S. law, which controls the disposition of plutonium produced in fuel of U.S. origin. Japan has built a special, lightly armed escort vessel to accompany plutonium-carrying freighters, but there are questions about whether this vessel as planned could protect against possible hijacking. The escort vessel is not under Japanese military jurisdiction because Japanese law prohibits the deployment of military forces beyond national borders.

Economic considerations do not appear to be the driving force behind Japan's decision to use plutonium for electrical power production. Japan has stated a desire to achieve independence from the import of fossil fuels. However, it is widely recognized that use of plutonium as fuel in light-water reactors is not likely to be justifiable on economic grounds for a long time because of the current low price of uranium. Consequently, economic uncertainties and growing public opposition to reprocessing and plutonium use for environmental, health, and security reasons may erode the Japanese plutonium program.[61]

As the U.S. and Soviet nuclear arsenals are reduced, world concern will focus on the stockpiling of plutonium in the civilian sector. This is likely to grow into a major issue by the time of the 1995 Nuclear Non-Proliferation Treaty extension conference. Nevertheless, the financial and institutional commitment to reprocessing in France and Britain is very great, and breaking away would be difficult.[62] In addition to the fact that huge amounts of capital have been invested in newly completed reprocessing facilities in the U.K. and France, the waste management programs of these countries have been designed and implemented on the assumption that spent fuel will be reprocessed. For example, in the U.K., the chemical composition of irradiated Magnox fuel is such that it cannot be stored for long periods in its water-filled storage basins without suffering significant degradation of the cladding. This problem of degradation of irradiated fuel placed in water basins could have been avoided if the U.K. had planned for general storage of this fuel in dry casks (instead of water basins). The British nuclear industry has readily admitted this fact: "With hindsight, dry storage would have been preferable to wet storage if adopt-

61 Berkhout et al. 1990; Sanger 1991.
62 Berkhout et al. 1990, pp. 541–543.

ed from the outset since it would have given more flexibility. . . and would have led to reduced radioactive discharges to the environment."[63] The U.K.'s Wylfa reactor station in Wales has been successfully using dry storage technology since 1971, but it is the only one. Presumably, such a decision for dry-cask storage can still be implemented, though at additional cost. Thus, in the U.K. the current need for reprocessing is not so much a technological imperative as a political decision regarding spent fuel management.

Fabrication of Weapons from Plutonium

PROCESSES[64]

Once the plutonium has been produced, it must be machined and shaped into use for a weapon. This form is generally called the "pit" or core of a fission weapon (or fission trigger of a thermonuclear weapon) — that part inside the spherical chemical high-explosives that detonate the device.

The principal component of a pit is plutonium-239, in the form of a sphere at the center. This spherical plutonium core is surrounded by uranium and beryllium tampers used to reflect neutrons back into the core. The pit may also contain a hollow core into which a tritium-deuterium gas mixture is automatically injected just before detonation (the tritium and deuterium are used to boost the yield of the weapon).

The principal industrial processes involved in the fabrication of nuclear warhead pits are metallurgical production and chemical processing activities with plutonium, uranium, and beryllium metals. Some conventional metals (such as stainless steel) are also used in the production of warhead pits. Required metallurgical operations include the reduction of plutonium oxide (PuO_2) to convert the plutonium to its metal form. This is followed by various treatments to produce plutonium ingots. The plutonium must then be formed, machined, and joined into the pit subcomponents. (Similar metallurgical operations are conducted with uranium, beryllium, and other more conventional metals used in pit fabrication.)

Chemical processing activities are essential for the recovery and purification of plutonium from scrap and from retired weapons, a high-priority capability, considering the expense of this material. Chemical processing

63 U.K. CEGB 1986, p. 2.
64 This section is largely based on Cochran et al. 1987b, pp. 82–84.

may also be used in the recovery of beryllium and the extraction of americium-241 from plutonium.

HAZARDS

Routine Hazards

In order to minimize the health risks presented by plutonium at a weapons plant, extensive safety measures must be implemented. One of these is the use of glove boxes for plutonium processing and fabrication. Glove boxes are specially enclosed units in which materials are manipulated through gloves installed in sockets on the wall of the enclosure. The gloves in glove boxes, however, can tear or may develop holes, thereby exposing workers to contamination.

In the plutonium pit manufacturing facility used in the U.S. (the Rocky Flats Plant near Denver, Colorado), some work areas are severely contaminated, and some production operations there make contamination control virtually impossible, according to a U.S. National Academy of Sciences (NAS) review of safety problems within the U.S. weapons complex.[65] This has led to the extensive use of respirators at the U.S. plant. The NAS review criticized the "respirator culture" at Rocky Flats, the feeling that "as long as workers wear respirators, it is unnecessary to seek to maintain a contamination-free work area." The review found that overreliance on respirators had several negative consequences, including the strain placed on the lungs and increased fatigue caused by their use. The most serious disadvantage noted, however, was the false sense of security, the feeling that a respirator would prevent any radioactive inhalation problems. In fact, the review noted that practices that rely on maintaining an uncontaminated work environment are on the order of 100 times more effective in protecting workers than relying on respirators to protect workers in a contaminated environment.[66]

Mismanagement can reduce or eliminate the effectiveness of safety measures. So, for example, at the U.S. Rocky Flats Plant, there have been numerous instances of accidental releases and spills of plutonium. At least one of these involved the backwards installation of air filters which resulted in the release of 26 curies of alpha-emitters to the atmosphere.[67]

---

65 U.S. National Academy of Sciences 1989, p. 57.

66 In one facility considered, committed lung dose dropped from more than 100 person-rem to 1.1 person-rem as a result of changes in work practices that reduced reliance on respirators for worker protection.

67 Coyle et al. 1988, p. 91.

Accidental Criticality

At the Rocky Flats plutonium pit production plant in the U.S., kilogram quantities of plutonium have accumulated in the plant's ductwork. In addition to increasing the risk of accidental release to the environment, such accumulation of unknown quantities of plutonium in unknown configurations poses the threat of accidental nuclear criticality.[68]

Nuclear criticality occurs when sufficient fissionable material is assembled in a small enough area to sustain a nuclear chain reaction in which each fission event in turn causes another fission event. The consequences of such an accident would not be comparable to a full-scale nuclear explosion,[69] but intense radiation would likely be produced, and environmental pollution as well as damage and injury could be significant.

Accidental criticality is known to have occurred within the U.S. nuclear weapons complex on at least eight occasions, sometimes with fatal consequences (there were two workers killed as a result of lethal acute doses) and sometimes with numerous significant exposures. Three of the events were plutonium criticalities, five were uranium, and all occurred in material in solution. Such events have typically produced radiation that is potentially lethal within a radius of about 10 meters. None of the events resulted in the explosive release of energy. For all the incidents, the energy release in the first few seconds was on the order of one kilowatt-hour, and the greatest energy release is estimated at 100 kilowatt-hours over a period of hours.[70]

Numerous instances of near criticality have also occurred. For example, in 1973 a burial trench at the U.S. Hanford facility had to be excavated because of concerns that the concentration of plutonium at the bottom of the trench (accumulated from the disposal of plutonium-contaminated liquids) might be great enough to cause a nuclear criticality.[71]

Plutonium-Induced Fires

Because plutonium metal reacts exothermically (releasing energy as heat) with air, and because plutonium is also a poor conductor of heat, it can be pyrophoric, meaning it has the ability to spontaneously ignite in air. Plu-

---

68 U.S. National Academy of Sciences 1989, p. 58.

69 Under accidental super-criticality conditions (i.e., where more than one fission results from each fission, causing the reaction to rapidly increase), the growing energy from the nuclear reaction would blow the material apart, stopping the reaction long before it could produce the energy released by a nuclear bomb.

70 U.S. National Academy of Sciences 1989, p. 117.

71 Lipshutz 1980, p. 131.

tonium in the form of lathe turnings is particularly susceptible to this hazard, and several serious fires in the U.S. nuclear weapons complex have been caused as a result of spontaneous plutonium ignition.[72]

Depending on the location of a plutonium pit fabrication facility, such fires could threaten the surrounding population, considering the large amounts of plutonium that could be vaporized and released to the atmosphere if such a fire breached a building structure.

Two of the most serious fires in the U.S. nuclear weapons complex occurred in 1957 and in 1969 at Rocky Flats, Colorado, a high risk plant because of typically high winds and the nearby, downwind location of a highly populated metropolitan area (Denver). In the 1957 fire, the burning of the air filters alone is presumed to have released somewhere between 10 and 230 kilograms of plutonium to the atmosphere. In the 1969 fire, about 1 ton of plutonium burned, monitoring devices were destroyed, and some filters were breached, releasing an unknown quantity of plutonium to the environment.[73] This fire, believed to be due to the spontaneous ignition of plutonium scrap, was one of the most costly industrial fires in U.S. history, causing upwards of $50 million in damages. Had the fire burned through the roof of the plant, hundreds of square miles of the Denver area might have been contaminated by plutonium aerosols.[74]

We turn now from discussing plutonium itself to a consideration of the wastes generated by its production and the associated hazards, a discussion that occupies the next four chapters.

72 U.S. National Academy of Sciences 1989, p. 117.

73 Good vs. Church, Church vs. Dow and Rockwell 1978.

74 Carter 1987, p. 66.

Chapter 3
Radioactive Wastes from Plutonium Production

REPROCESSING OPERATIONS GENERATE large amounts of waste. This includes high-level liquid wastes, which are typically stored in double-shell underground steel tanks, as well as intermediate-level and low-level liquid and solid wastes, which have been discharged to cribs, ponds, rivers, and landfills. Radioactive gases are also generated and discharged to the air after greater or lesser degrees of treatment and filtration. (These waste categories are discussed below.)

Thus, virtually any country which has plutonium-based nuclear weapons or civilian fuel cycle has produced at least some amount of waste which must be managed over the short and long-term.[1] High-level waste is particularly troublesome in this regard, as are plutonium-contaminated materials resulting from the handling of plutonium in separation, processing, and manufacturing activities.

Much of the actual and potential damage associated with worldwide plutonium comes not just from the plutonium itself, but from the management of these wastes. For this reason, this study devotes considerable attention to the management of high-level waste. After giving an overview of all wastes produced, this chapter will discuss the nature of high-level waste, the types and amounts of it in various countries around the world, and some of the short-term environmental and occupational problems posed by this waste. In the subsequent chapters we will discuss in greater detail the risk of catastrophic explosions in containers holding this waste. The explosion of a high-level waste tank in the former Soviet Union was one of the world's

---

1 As noted previously, some countries have shipped their irradiated fuel for reprocessing abroad, and thus may not have high-level waste currently stored within their boundaries. France and the U.K., for example, have processed spent fuel for many countries.

worst nuclear accidents and illustrates the grave risks of dealing with these wastes. Chapter 6 will discuss in detail the particular problems associated with the long-term management and disposal of high-level waste.

Wastes and Hazards from Reprocessing

WASTE CATEGORIES

Radioactive waste categories are defined differently in different countries. Most countries, however, have the categories "high-level," "intermediate-level," and "low-level" wastes, which are organized roughly according to level of radioactivity. Such definitions are not necessarily consistent among countries, however, as can be seen in Table 3.1, which shows how the major waste categories are defined in the U.K., Russia and the U.S.

The U.S. is different from most countries in that its waste categories are defined not so much by the level of radioactivity they contain but by the process that produced them. For example, all spent fuel and waste from the solvent extraction reprocessing of spent fuel are considered high-level. Virtually all other wastes — from reactor operation, for example — are considered "low-level," even though some "low-level" waste can be quite radioactive and in some cases exceeds the radioactivity of some "high-level" wastes.

Aside from the waste categories outlined above, many countries also have a separate category for what are called "plutonium-contaminated materials" (PCMs). These are wastes contaminated with significant amounts of plutonium. Such wastes result from plutonium separation and processing activities like those associated with weapons production. In the U.S. they are called "transuranic wastes," due to the fact that plutonium is a transuranic material.

For the purposes of this chapter, the primary focus will be on the liquid wastes that are produced from separating plutonium at a reprocessing plant. This waste is usually categorized as "high-level", at least until further processing, which may separate it into high- and low-activity portions.

REPROCESSING WASTES[2]

The major source of reprocessing wastes is the aqueous waste stream from the solvent extraction process. This waste stream contains more

2 Uses the Purex process as an example, as described in U.S. DOE 1982, pp. 3–7 to 3–11, and A.20.

Table 3.1. Waste category definitions

WASTE CATEGORY	U.K.[a]	RUSSIA[b] (liquid waste)	U.S.[c]
High-level	> 15 W/m^3 [d]	> 37,000 GBq./m.^{3}	spent fuel or reprocessing waste
Intermediate-level	in between	in between	not applicable[e]
Low-level	< 12 GBq./m.^{3} of gamma or beta (or < 4 GBq./m.^{3} of alpha)	< 37 GBq/m.^{3}	everything not classified as "high-level"[f]

(a) Large 1992b.
(b) Soyfer et al. 1992.
(c) Discussed in Makhijani and Saleska 1992.
(d) Heat generation rate as a result of radioactivity.
(e) There is no "intermediate-level" waste category in the U.S.
(f) Low-level waste is not defined in the U.S. by activity level; all waste in the fuel cycle that is not spent fuel or wastes resulting from the reprocessing of spent fuel is categorized as low-level waste (except for uranium mill tailings and some wastes contaminated with plutonium and other transuranic elements.).

than 99 percent of the fission products (including cesium and strontium), about 10 percent of the total neptunium, and trace quantities of plutonium and uranium. This waste is sometimes treated and concentrated before being discharged to tanks for storage. If this waste is to be stored in carbon steel tanks, it must first be neutralized, since it is highly acidic and carbon steel is easily corroded. Except at the U.S. facilities at Hanford, Savannah River, and West Valley, most tanks for waste storage worldwide are believed to be built to hold acidic wastes (for example, constructed with stainless steel).

Other lesser contaminated liquid wastes include process and scrubber wastes, steam condensates, cooling water from heat exchangers, and chemical sewer waste.

Relatively small (in comparison to the liquid waste volumes) amounts of transuranic and low-level solid radioactive wastes are typically generated from reprocessing as well and consist primarily of failed and unusable production equipment, tools, laboratory equipment, and other materials. The low-level solid wastes are typically disposed of by burial, and the plu-

Table 3.2. Waste generation and radioactivity from reprocessing, in terms of selected radionuclides , N-reactor, Hanford, USA [a]
(per 1,000 tons of typical irradiated fuel)

WASTE FORM	VOLUME	RADIOACTIVITY, CURIES (selected radionuclides)
HIGH-LEVEL LIQUID	$363 - 570 \times 10^3$ gallons [b]	[c]
Strontium-90		3.3×10^6
Iodine-129		0.95
Cesium-137		3.9×10^6
Plutonium-239,240		2.6×10^2
Americium-241		2.4×10^4
LOW-LEVEL LIQUIDS [d]	1.9×10^9 gallons	
Tritium		4.0×10^3
Carbon-14		2.5
Strontium-90		0.48
Cesium-137		5.5
Plutonium-239,240		1.2
LOW-LEVEL SOLIDS [e]	218 cubic meters	not available
DISCHARGES TO AIR [f]	not applicable	
Tritium		2.3×10^3
Krypton-85		9.5×10^5

(a) Based on estimated waste generation from processing the remaining spent fuel (2,100 tons worth) stored at the U.S. Hanford Nuclear Reservation at the Purex plant.
(b) Lower estimate from U.S. DOE 1986, p. 5.6; upper estimate from calculations based on the assumption by Strode et al. 1988 (p. 11) of 587 gallons per ton of uranium
(c) Radioactive inventories in high-level waste are calculated on the basis of data in U.S. DOE 1987, p. A.18. These estimates ignore short-lived nuclides.
(d) U.S. DOE 1986, p. 5.6. Includes discharges to both ponds and cribs.
(e) U.S. DOE 1986, p. 5.6.
(f) U.S. DOE 1986, p. 5.6.

tonium-contaminated wastes are today placed in retrievable storage in preparation for their eventual disposal.[3]

Large volumes of radioactive waste will also be generated when reprocessing plants are decommissioned.

Table 3.2 lists the average amount of various wastes generated by processing irradiated fuel from the N-reactor at the Hanford site in the U.S. It

---

3 In the U.S., the DOE plans to dispose of transuranic wastes at a repository (known as the Waste Isolation Pilot Plant, or WIPP) constructed but not yet opened near Carlsbad, New Mexico. WIPP is encountering numerous environmental problems, most notably the leakage of water into what was intended to be a dry repository. In the past the U.S. disposed of these wastes by shallow land burial and sea-dumping.

should be noted that some of these figures — and treatment procedures used on the waste — are highly dependent on variable factors such as burn-up levels. In addition to their radioactive content, many of the wastes generated by reprocessing are chemically hazardous as well. Some of these hazardous components include n-butyl alcohol, acetone, and ammonia.[4]

RELEASES DURING REPROCESSING

Among the problems associated with reprocessing are releases of radioactive materials during the process, which is distinct from the long-term problem of managing the wastes which result from it. Of particular note are the gaseous fission and activation products, which, if not carefully managed, can be easily released to the atmosphere. The radioactive gases of concern include xenon, tritium, carbon-14, the radioiodines, and krypton-85. Removal of the radioiodines is particularly important because of their toxicity. (In the case of iodine-131, which has a half-life of only 8 days, aging of the fuel for a few months before reprocessing will allow most of it to decay away.) About 1 percent of the iodine in the spent fuel is volatilized during decladding and a larger fraction during dissolution. Some remains dissolved and must be removed before entering the solvent extraction phase because it would cause problems in the process. The volatilized iodine is typically removed by routing the gas to iodine absorbers before gases are released to the atmosphere.[5]

If insufficient time is allowed for iodine-131 decay or if it is not removed (and in the past it sometimes has not been), potentially huge health risks can result. For example, in the late 1940s in the U.S., large amounts of iodine-131 and other gaseous radionuclides were released during reprocessing runs of un-aged spent fuel at the Atomic Energy Commission's Hanford plant. Perhaps the most infamous of these, known as the Green Run, was a deliberate, experimental release of about 11,000 curies of iodine-131 on December 2 and 3, 1949.[6] This resulted from the reprocessing of a fresh or "green" batch of fuel (only 16 days out of the reactor), in which operators disconnected stack filters to maximize the amount of iodine released.[7] Although some details of this experiment remain classified to this day, including the reason for the release, the official public explanation for this release is that it was intended to facili-

4   These RCRA-regulated waste streams are discussed in Attachment B of Natural Resources Defense Council et al. 1990.

5   Benedict et al. 1981, p. 481.

6   Dr. Maurice Robkin, cited in Thomas 1992.

7   Based on official documents reviewed in Thomas 1990, p. 6.

tate the development of "a monitoring methodology for intelligence efforts regarding the emerging Soviet nuclear program."[8] All in all, according to government documents that are now public, the Hanford reprocessing plants released some 515,000 curies of iodine-131 in the years 1944–1949.[9]

In 1990, a panel (independent of the DOE) called the "Technical Steering Panel" (TSP) reported its estimates of doses received due to the iodine-131 releases and their consequences. The panel concluded that an infant born in 1945 and living across the river from Hanford could have received a dose to the thyroid of as much as 29 Grays (2,900 rads), and that in all, 13,500 people received thyroid doses of 0.33 Gray (33 rads) or more.[10] This confirmed results of an earlier panel which found that residents near Hanford in the 1940s received higher doses to the thyroid from radioiodine than people living in the immediate vicinity of the 1986 Chernobyl reactor explosion. According to this estimate, more than 30,000 children may have increased their chances of getting thyroid cancer 5- to 15-fold.[11]

The noble gases krypton and xenon are also of concern, although less so than the radioiodines. Krypton-85, with a half-life of about 10 years, is the more worrisome, since after a year the radioactivity of the xenon would be negligible. Krypton, however, is especially difficult to remove and is often released to the atmosphere. With proper equipment the krypton-85 could be held up to prevent its release, although this is very expensive; even the most modern reprocessing plants such as the THORP plant in the U.K. (now scheduled to open in late 1993 or 1994) do not plan on krypton-85 retention.[12]

EXPLOSIONS DURING REPROCESSING

Uranyl or plutonyl nitrates in contact with tributyl phosphate solutions, a mixture sometimes called "red oil," can explode. One such incident

8 Letter from Michael Lawrence, DOE/Richland Manager, to Representative John Dingell, Chair of Energy and Commerce Committee, April 4, 1986, as quoted in Thomas 1990, p. 6.

9 As reported in Thomas 1990, p. 5. Note that this figure is a sum of the radioactivity as measured or estimated at the time of release. Because of iodine-131's 8-day half-life, the total amount in the environment decreases rapidly and essentially none of this isotope remains in the environment from Hanford releases. In contrast, iodine-129 from past releases has a half-life of about 16 million years and still exists practically undiminished.

10 TSP Chairman John Till, as reported in Thomas 1990, p. 10.

11 Alvarez and Makhijani 1988, p. 46.

12 Large 1992b.

occurred at Savannah River Laboratory in January 1953. A batch solution of uranyl nitrate and nitric acid was being concentrated in an evaporator. Approximately 80 gallons of tributyl phosphate had been fed to the evaporator. As the evaporation cycle was nearing completion, an explosion occurred, demolishing the evaporator and destroying or heavily damaging the roof and two sides of the building. An investigation of the incident revealed that the bubble cap trays (through which the evaporator vented) had become partially plugged, leading to a significant difference in pressure between the boiling and condensing regions of the evaporator. The explosion was determined to be a vapor phase deflagration of butanol and other flammable organic gases generated by the rapid hydrolysis of tributyl phosphate.[13]

The second incident occurred at Hanford in July 1953, during testing of a uranium concentrator. This unit was designed to concentrate uranyl nitrate solutions. During the first test of this concentrator, nitrogen dioxide (NO_2) fumes erupted from the condenser vent with a hissing noise. The initial force of the reaction was sufficient to lift the concentrator about a meter off the floor. An investigation of the incident revealed that the cause of the accident was inadequate instrument control and large quantities (in excess of the removal capacity) of organic compounds that had been fed to the evaporator.[14]

Another incident occurred at Oak Ridge in 1959 with a solution similar in composition to those handled at Hanford and Savannah River. An explosion occurred in a radiochemical plant evaporator that was concentrating a nitric acid solution of plutonium nitrate possibly contaminated by organic solvents.[15]

High-Level Waste from Reprocessing

GENERAL BACKGROUND

High-level waste from reprocessing typically refers to the aqueous waste stream from the first-cycle solvent extraction column used to remove plutonium, uranium, and sometimes other actinides from spent nuclear fuel. Coming out of a reprocessing facility, as mentioned earlier, this waste stream contains 99 percent of the fission products, about 10 percent of the total neptunium (assuming the system is designed to separate most of

13 Tomlinson 1953.

14 Sege 1953.

15 Nuclear Safety 1960.

the neptunium, as many are), as well as a smaller percentage of the pluto-nium and uranium due to inefficiencies in extraction. This stream also contains most of the nitric acid used in the solvent extraction system, but much of this is usually recovered for re-use in the system.[16]

This stream of waste is sometimes further processed. Such processing can include evaporation to concentrate the waste, the removal of some radionuclides (such as cesium-137 and/or strontium-90, which generate a substantial amount of heat by radioactive decay), the neutralization of the acidic waste, or the calcining of it to convert it to a solid form. This waste from the solvent extraction process is also sometimes mixed with wastes from other parts of the process, such as decladding wastes (which result from the removal of the cladding from the spent fuel before it is dis-solved).[17] In some cases the decladding wastes are stored separately.[18]

Further processing of wastes can change the waste form and can redis-tribute the radioactivity (although the total radioactivity present in the waste is unaffected by any processing and is governed only by the processes of radioactive decay).

MANAGEMENT AND STORAGE OF HIGH-LEVEL WASTE

Reprocessing invariably results in a large increase in the volume of radioactive materials to be managed. In general, high-level wastes in their liquid form are a highly hazardous, difficult-to-manage mixture of highly radioactive materials and toxic chemicals.

In the early years, this highly radioactive and toxic waste was sometimes discharged directly to the ground and surface waters. Perhaps the most egre-

---

16 Based on U.S. DOE 1982, p. 3.7.

17 This depends, to some extent, on the decladding mechanism used in the facility. Many places (such as the U.K.'s Sellafield facility and the U.S.'s now-closed West Valley reprocessing facility) have used a mechanical decladding process. At West Valley, the spent fuel was sheared into pieces, and the exposed irradiated uranium fuel leached from the cladding hulls in a nitric acid bath. The cladding of Magnox fuels in the U.K. are peeled off by forcing the fuel through a collet. The Purex plant at Hanford, by contrast, uses an entirely chemical process (called the "Zirflex" process) in which the Zircalloy cladding is dissolved in an ammonium fluoride - ammonium nitrate bath. (U.S. DOE 1982, p. 3.5.)

18 This appears to be the case in the U.K., for example, where the remains of prelimi-nary Magnox fuel decladding operations at individual reactors are mixed with water and stored in silos. At the U.K.'s reprocessing station in Sellafield, the cladding residues are compacted and immobilized in a cement grout for long-term storage. Prior to 1982, much of the U.K.'s Magnox cladding was dumped at sea. (Carter 1987, p. 237; Large 1992b.)

Figure 3.1. Chemical separations area (200H), Savannah River Plant, Aiken, South Carolina, USA, 1982. Photo by U.S. Department of Energy.

gious example of this occurred in the Soviet Union in the late 1940s and early 1950s. Waste from the Chelyabinsk-40 (now called Chelyabinsk-65) weapons complex in the Southern Ural Mountains was discharged directly into the nearby Techa River and then, after discovery of excess radioactivity as far away as the Arctic Ocean, into nearby Karachay Lake, which had no outlet.[19]

Since that time, and in most countries where reprocessing has taken place, these wastes have been stored in large tanks. The storage of high-level wastes in tanks was first done at the Hanford facility in the U.S., where these wastes were produced on a large scale in the mid-to-late 1940s. Originally considered an interim storage method to be used only until a long-term solution could be found, this method has become the *de facto* standard storage practice in the U.S. In some places, the wastes are solidified through a process called vitrification and then stored in solid form in anticipation of a long-term disposal method becoming available.

Capacities of high-level waste storage tanks can range to well over a million gallons, and there is a wide variety of designs. The highly radioactive waste can be stored in the same acidic form in which it was generat-

19 Cochran and Norris 1992.

ed in the reprocessing plant if the storage tank is constructed of a relatively non-corrodable substance such as stainless steel. This method is now used at most modern reprocessing plants around the world. In the U.S. plants at Savannah River and Hanford, however, the waste was first neutralized with sodium hydroxide, which allowed storage in tanks constructed of less costly carbon steel. This neutralization resulted in the separation of waste into a sludge layer at the bottom, which contains almost all the radioactivity except for cesium-137. This last remains in solution in the liquid supernate above the sludge layer. Aged wastes at Savannah River were concentrated by evaporation. This resulted in the conversion of much of the supernate portion of the wastes into crystallized salts.

In general, there are important differences in practices at the older military facilities (such as the U.S. reprocessing plants at Savannah River and Hanford) and the newer, commercially oriented ones (such as at La Hague in France and THORP in the U.K.). Since waste tanks are expensive and waste processing costs can be significant, there is a strong incentive in a commercially oriented operation to minimize waste volumes and to quickly convert wastes to solid form for easier storage. These incentives were much less relevant for older U.S. and Soviet plants where military considerations were predominant.

These reasons partially account for the sometimes great differences in chemical composition and volume of wastes at various sites around the world. The potential problems that can arise in the management and storage of these wastes are dependent on the practices followed, so the types and magnitudes of risk will vary as well.

The locations and amounts of these wastes are discussed in the section below; some of the general environmental problems that can arise from their management are discussed in the subsequent section.

Locations and Amounts of High-Level Waste

The locations and amounts of high-level waste stored will be reviewed here on a country-by-country basis, along with a brief description, when available, of the country's high-level waste management practices.

UNITED STATES

The total amount of high-level radioactive waste stored in the U.S. as of the beginning of 1991 as a result of defense reprocessing activities con-

Figure 3.2. High-level radioactive waste tanks under construction, Hanford Nuclear Reservation, Richland, Washington, USA, 1984. Photo by Robert Del Tredici.

tains about one billion curies, a figure which has been corrected for radioactive decay.[20] This includes about 245 million curies each of strontium-90 and cesium-137.

Hanford and Savannah River

About 560 million curies of waste are stored in 50 tanks at the Savannah River site in South Carolina,[21] and 224 million curies in 177 tanks at Hanford; in addition, about 86 million curies of strontium-90 and cesium-137 have been separated from the Hanford wastes and placed separately in double-walled capsules stored in a water basin.[22]

The Hanford tank waste occupies almost twice the volume but has less radioactivity than the Savannah River tank waste. The average concentration of radioactivity in the Savannah River tank waste is about 4,300 curies per cubic meter, whereas at Hanford it is about 890 curies per cubic meter. This is in part due to the fact that large portions of the radioactivi-

20 U.S. DOE 1991b, p. 47.

21 One tank of the 51 at Savannah River has been emptied.

22 There are 1345 cesium capsules (occupying a total of 2.5 cubic meters) and 597 strontium capsules (1 cubic meter). (U.S. DOE 1991b, p. 40.)

ty (in the form of cesium-137 and strontium-90) of the Hanford waste has been removed from the tanks and stored in separate capsules, and is also due to the fact that many of the wastes at Hanford were generated by older reprocessing methods (such as the reduction-oxidation, or Redox, process), which were less efficient and produced a greater volume of waste per unit of radioactivity. The Savannah River reprocessing facilities have always used the more volume-efficient Purex process.

The DOE's original waste tanks were constructed as "single-shell" tanks, consisting of a single-walled carbon steel shell with an outer concrete envelope. The outer envelope has a steel pan to catch leaks from the primary containment. Many of the single shell tanks developed leaks (9 of 16 at Savannah River and up to 66 of 149 at Hanford), so a "double-shell" model was developed consisting of two concentric steel cylinders.

The chemical composition of tank waste consists primarily of the radioactive constituents left over from the solvent extraction process, which are in solution as nitrates and nitrites, and various other waste products such as sodium compounds resulting from the neutralization of these wastes.

In addition, the tanks contain a number of other chemicals which were either added on an *ad hoc* basis or produced as a result of radiolytic or chemical decomposition. These chemicals have included organic complexants and the chelating agents EDTA and HEDTA. These organic constituents were used to remove cesium-137, strontium-90, and other fission products from the waste. In the case of Hanford, ferrocyanide was added to about 20 tanks in the 1950s for the purpose of precipitating cesium-137.[23] Measurements of organics in Hanford tanks indicate concentrations as high as 500 grams of carbon per gallon of waste.[24] Organic compounds increase the combustibility of the mix, as explained in Chapter 5.

At Savannah River, sodium tetraphenylborate (STPB) is used to precipitate cesium. Decomposition of STPB has produced toxic substances such as benzene, nitrobenzene, phenol, and biphenyl in the waste, concentrated most in two tanks used in the precipitation process.[25]

Idaho National Engineering Laboratory and West Valley

In addition to Savannah River and Hanford, the Idaho National Engineering Laboratory has about 64 million curies of waste split between an acidic liquid waste form and a calcined solid form; and the now-closed

23 U.S. DOE 1990b.
24 Van Tuyl 1983.
25 Du Pont 1988, p. 5–53.

commercial reprocessing plant near West Valley, New York has two tanks containing about 27 million curies of alkaline and acid wastes.[26]

Reprocessing activities at the Idaho National Engineering Laboratory were conducted at the Idaho Chemical Processing Plant (ICPP) until the spring of 1992, when it was shut down. These activities were not directed at recovering plutonium but rather at recovering the highly-enriched uranium in naval reactor fuels used to power ships and submarines. Because they are highly enriched, such fuels contain little plutonium.[27] However, the waste management issues are essentially the same, so for this reason the Idaho facility is covered here.

High-level wastes at INEL are first stored in interim underground storage until they are sent to the New Waste Calcining Facility (NWCF). There are, however, 1.5 million gallons of sodium-contaminated waste that cannot be calcined.[28]

There are a total of 15 high-level waste storage tanks at INEL: eleven 300,000-gallon tanks and four 18,400-gallon tanks, all housed in underground reinforced concrete vaults. Some of these are not cooled, but most have cooling coils. If necessary, additional storage can be provided by four 30,000-gallon cooled underground tanks that are normally kept empty and used only by special authorization for *ad hoc*, nonroutine processing. All of these tanks are constructed of stainless steel to allow direct storage of acid waste. The tanks are interconnected via pipelines to allow transfers between tanks. If all the tanks available for a certain waste blend are full, reprocessing of fuel producing that type of waste must be terminated until some of the waste volume can be solidified in the calciner (NWCF).[29]

There are two tanks containing high-level wastes at the former West Valley reprocessing facility. One contains acidic wastes, and the other contains neutralized wastes. There have been a number of serious environmental problems from waste disposal at this site, including water contamination arising out of low-level waste disposal. While the plant itself cost only about $32 million and operated for six years (1966–1972), the total costs for dealing with the radioactive wastes arising from its operations will be over $3 billion. The plant was supposed to be a private com-

---

26 U.S. DOE 1991b, p. 54.

27 In highly enriched uranium fuel, there is little uranium-238 present, and therefore little to be converted to plutonium-239.

28 Snake River Alliance Bulletin 1992.

29 WIN 1990, Section 4.2.2.

mercial venture, but these waste disposal costs are being borne by the tax-payers of New York State and the United States as a whole.

SOVIET UNION

The former Soviet Union continues to keep secret essential data on radioactive wastes from plutonium production. This is unacceptable, especially in view of the damage to human health and environmental quality caused by past Soviet plutonium production and the fact that plutonium is still being produced there.

Chelyabinsk-65

High-level liquid wastes produced from Chelyabinsk-65 reprocessing facilities are stored in tanks. According to Soviet sources, there are 99 stainless steel tanks at Chelyabinsk-65 housed inside concrete, and the tank capacity is 300 cubic meters each. A total of 823 million curies of waste is reportedly contained in these tanks.[30] We do not have the figures for the volume of waste that is stored in the tanks, but this must be less than 30,000 cubic meters, since that is the maximum capacity of the tanks for the storage of these wastes. Of course, the quantities of high-level waste discharged directly into the Techa River (containing a reported 2.75 million curies of beta emitters) and into Lake Karachay (including 120 million curies of cesium-137 and strontium-90) are not included in these volume and radioactivity figures.[31]

In 1987, a pilot-scale 500-liter-per-hour vitrification plant went into operation at the Chelyabinsk-65 site. The Chelyabinsk-65 vitrification process incorporates the radionuclides into a phosphate glass. The resultant glass blocks are packaged in metal containers and placed in surface storage facilities where they are cooled via a forced air system. This mode of storage is expected to be maintained for 20 to 30 years, after which the Soviet plan was to bury the waste in a deep underground repository, possibly in the Ural mountains.[32] In May 1992 it was reported that 60 million curies had been vitrified. The currect concentration of radioactivity achieved is 400 curies per liter, up from the earlier level of 100 curies per liter.[33]

---

30 Soyfer et al. 1992. Cochran and Norris 1992, p. 45, report that Chelyabinsk-65 has "approximately 60" single-walled steel tanks.

31 Soyfer et al. 1992, p. 6. Cesium-137 makes up about 100 million curies of the total 120 million curies reportedly contained in lake Karachay.

32 Cochran and Norris 1992, p. 47.

33 Oleg Bukharin's notes of a May 28, 1992 meeting in Moscow, cited in Cochran and Norris 1992, p. 46.

Tomsk-7

There has been a considerable amount of plutonium production at a reprocessing plant located near Tomsk in Siberia. Accounts of waste management there are vague or conflicting. One source states that there are no waste storage tanks there, that about 100 million curies of medium- and high-level wastes, including neptunium, strontium-90, and cesium-137, have been discharged into two artificial reservoirs near Tomsk, and that 1 billion curies of liquid wastes have been injected underground.[34] According to a Russian newspaper account, about 127,000 tons of solid and about 33 million cubic meters of liquid radioactive wastes have been collected in underground storage facilities. This source says, in addition, that radioactive wastes of unknown quantity and concentration have been pumped into sandy beds at a depth of 220-360 meters. These beds, located some 10–20 kilometers from the Tom River, are said to be covered with water-resistant clay strata; over the region as a whole, however, these clay layers can thin out.[35]

Whatever the reliability of these accounts, the estimates of plutonium production in the former Soviet Union indicate that the amount of plutonium produced at the Tomsk-7 plant was of the same order of magnitude as that at Chelyabinsk. Thus, we would expect that there are well over 500 million curies of high-level radioactive waste deposited at Tomsk-7. Based on the plutonium production estimates for Tomsk-7 reported in the last chapter, we estimate that some 314 million curies of strontium-90 and cesium-137 must have been created; the undecayed remainder of this is presumably either in storage or dispersed into the environment.[36]

Krasnoyarsk-26

Based on an estimated 30 tons of plutonium-equivalent production, there has been an estimated 170 million (not decay-corrected) curies of strontium-90 and cesium-137 generated at the underground chemical separation facility at Krasnoyarsk-26. As at Tomsk-7, waste quantities and disposal practices have been closely guarded secrets. However, Alexander Bolsunovsky, senior researcher at the local branch of the Institute of Biophysics and director of the Krasnoyarsk Ecological Center, reported in

34 Penyagin 1991.

35 Cited by Cochran and Norris 1992, p. 51. Main article cited, which quotes from two additional sources, including an official one by specialists in Tomsk, is V. Kostyukovskiy et al., "Secrets of a Closed City," Moscow Izvestiya, Union Edition, August 2, 1991 (translated in JPRS-TEN-91-018, October 11, 1991, pp. 71–72).

36 Based on approximately 3 curies each of cesium-137 and strontium-90 created for each gram of plutonium produced.

1992 that high-level liquid wastes from reprocessing are injected underground to a depth of 270 meters. This same source cites a newspaper article stating that in addition to being injected underground, high-level liquid wastes are placed in concrete tanks underground.[37]

UNITED KINGDOM

High-level waste is generated by reprocessing operations at the Sellafield (formerly Windscale) nuclear facility, run by British Nuclear Fuels (BNFL). The waste is first concentrated by evaporation and stored in stainless steel tanks in acidic form. High-level waste storage tanks include eight cylindrical tanks of about 60 cubic meters capacity each, arranged horizontally, and seven more recent tanks, which are vertical cylinders with a capacity of about 160 cubic meters. These later tanks have significant cooling capability of up to 2 megawatts (about 13 watts per liter). BNFL refers to these as High Activity Storage Tanks (HAST).[38]

Official British sources report a total of 1,430 cubic meters of high-level waste as of the beginning of 1987, containing over 800 million curies.[39] This means that British high-level wastes contain about 570,000 curies per cubic meter of waste. This concentration is a couple of hundred times greater than the wastes in the U.S.

There are two main reasons for the difference. For one thing, U.S. military wastes were neutralized (except at INEL), and other substances were added to the tanks over the years. In contrast, British acidic wastes are further concentrated after discharge from the reprocessing plant. Another major reason for the difference is that the wastes in Britain result from reprocessing a mixture of civilian and military irradiated fuel, whereas all of the high-level wastes at the Savannah River Site and most of the wastes at Hanford are the result of reprocessing fuel irradiated for military plutonium production. As explained in the section on nuclear reactors in Chapter 2, civilian fuel is generally far more highly irradiated, typically by a factor of ten more (although this is not so for Magnox reactors in the U.K. due to the special physical limitations of the design). Thus, the separation of plutonium from it yields more highly radioactive wastes.

37 Bolsunovsky 1992. Newspaper citation was Ekologia Krasnoyaria, No. 12, Dec. 1991–Jan. 1992.

38 From Large 1992a.

39 U.K. Nirex 1988.

FRANCE

Most of the high-level waste from French reprocessing operations is stored in double-walled, stainless steel tanks cooled by an electrical refrigeration system;[40] the rest is vitrified, a more recent process. According to data published by Cogema (Compagnie Générale des Matières Nucléaires), the company that operates the La Hague reprocessing plant, each ton of irradiated fuel from a light-water nuclear power reactor contains about 850,000 curies of radioactivity at the time that the fuel is reprocessed, about three years after it is removed from the reactor. Reprocessing generates about half a cubic meter of high-level radioactive waste per ton. This waste contains almost all the fission products and a small fraction of the alpha emitters, including plutonium, from the irradiated fuel. This means that at the time of discharge from the reprocessing plant, the high-level waste contains about 1.7 million curies per cubic meter. This is a very high level of radioactivity, hundreds of times larger than the average concentration in U.S. high-level wastes. The difference is largely due to the much longer periods of irradiation of the commercial fuels reprocessed at La Hague, the more concentrated French wastes, and also the fact that the waste is, on average, not as old, and therefore has had less time to decay.[41]

The radioactivity per cubic meter would be expected to decline to less than half over another five-year period, to about 0.8 million curies per cubic meter, due to the decay of short-lived fission products (such as ruthenium-106). Assuming that the waste is on average five years old, we may take this as the typical level of radioactivity in tanks containing high-level waste from light-water reactor reprocessing.

In the initial years (1966 to 1976), La Hague reprocessed fuel from graphite-moderated reactors.[42] This fuel was irradiated to a far lower degree, typically 4,000 megawatt-days per ton. This meant that the radioactivity at the time of reprocessing was far lower than that for light-water reactor irradiated fuel, amounting to about 30,000 curies per ton. Assuming the same volume of high-level waste per ton of reprocessed fuel

---

40 Barrillot 1991, Part II, Sect. 2.24 .

41 Cogema undated-a, pp. 11 and 14. The assumption that Cogema used in calculating these figures for radioactivity is that the fuel would be irradiated to 33,000 megawatt-days per ton.

42 La Hague has reprocessed fuel from both graphite-moderated and light-water reactors since 1976.

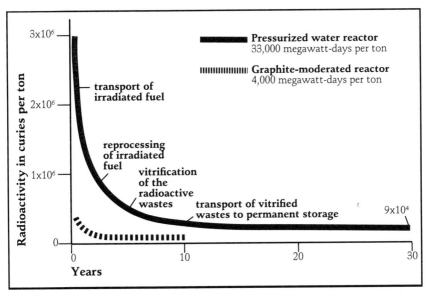

Figure 3.3. Radioactive decay in French high-level waste, shown for two burn-up levels, in curies per ton. Source: Cogema.

(half a cubic meter) and a decay by one-third to one-half after five years of storage, the typical high-level waste from graphite-moderated fuel would contain about 30,000 to 40,000 curies per cubic meter of waste.[43]

Overall, about 5,500 tons of irradiated fuel had been reprocessed at La Hague through the end of 1989,[44] about 2,600 tons from light-water reactors[45] and the rest from graphite-moderated reactors. Thus, we would estimate that the total amount of radioactivity in the tanks would be on the order of about 1 billion curies. We assume that most of this waste is still in liquid form. A vitrification plant went on line at La Hague in 1989; a second one is due to start up in July 1993.[46]

Cogema has also reprocessed fuel for nuclear weapons at its plant at Marcoule, in the south of France, since 1958. Plutonium production at Marcoule has amounted to about 16.9 tons, not including an estimated 6 tons of plutonium dedicated to military purposes (see Table 2.1 in Chap-

43 Cogema undated-a, pp. 11 and 14. These figures for waste from reprocessing of fuel from graphite-moderated reactors are approximate.

44 La Gazette Nucléaire 1989, p. 21.

45 Cogema 1989, p. 5.

46 Odell 1992.

ter 2). With the assumption of 6 curies of strontium-90 and cesium-137 generated per gram of plutonium produced, we arrive at an estimated inventory of about 130 million curies of cesium and strontium in the high-level waste in liquid or vitrified form at Marcoule.[47]

We have no exact figures on the partition of the radioactivity between liquid and vitrified wastes at Marcoule, but we assume that most of the waste is vitrified since a commercial-scale vitrification plant has been in operation since 1978.[48] (France was the first country to develop a commercial-scale vitrification facility for the solidification of liquid high-level wastes. A pilot vitrification plant was operated at Marcoule between 1969 and 1974, producing 12 tons of glass (containing 5 million curies) from 25 cubic meters of high-level liquid waste. This was followed by the full-scale vitrification plant there beginning in June of 1978.)[49]

CHINA

We have no data on Chinese high-level wastes, since China continues to be a very secretive country on nuclear military issues, even when they relate to environment, health, and safety. Our estimate of plutonium production for military purposes in China is between 1.25 and 2.5 tons. This gives an estimate of 7.5 to 16 million curies of non-decay-corrected strontium-90 and cesium-137 originally in the waste.

SUMMARY AND EXPLANATION OF VARIATIONS

The amounts of high-level waste and types of storage are summarized in Table 3.3 on a country-by-country basis.

The radioactivity of high-level waste per ton of plutonium produced varies somewhat due to the difference in irradiation of the fuel and the types of reactors that have been used. The range spans 6.5 million curies per ton of plutonium produced (Hanford, U.S.), 12 million (Savannah River, U.S.), 14 million (West Valley, U.S.), and 16 million (Sellafield, U.K.). In the early years at Hanford, the processes that were used to produce plutonium generated larger quantities of waste per unit of plutonium production.

The radioactivity per unit volume of liquid high-level waste is much more varied due to differences in irradiation of the fuel. Also, in the case of the U.S., large increases in the volume of waste have resulted from neutralization of

47 Note that these figures have not been corrected for radioactive decay.

48 Cogema undated-b.

49 Barrillot 1991, p. 36.

Table 3.3. High-level waste stored, by country

COUNTRY/ FACILITY	WASTE STORAGE FACILITIES Number (approx. capacity)	Waste Type	WASTE AMOUNTS Volume (m3)	Radioactivity (in millions of curies) (Sr-90+Cs-137) [a]	Total
RUSSIA					
Chelyabinsk-65	60-99 (300 m³)	acidic	<30,000 [b]	245	(945) [c]
		vitrified (glass)	162 tons		
Tomsk-7		acidic?		314	(N/A)
Krasnoyarsk-26		acidic?		170	(N/A)
UNITED STATES					
Hanford	149 (million gal)	alkaline ⎫	254,000	102	(224) [d]
	28 (million gal)	alkaline ⎭			
	1,942 (1.8 liter)	calcined capsules	3.5	86	(168) [d]
Savannah River	51 (million gal)	alkaline	132,000	262	(562) [d]
INEL	11 (300,000 gal)	acidic ⎫	8,500	3.6	(7.5) [d]
	4 (18,000 gal)	acidic ⎭			
		calcined solid	3,500	28	(56) [d]
West Valley	1	acidic	50	0.87	(1.7) [d]
	1	alkaline	1,136	12.8	(25.6) [d]
UNITED KINGDOM					
Sellafield	8 (70 m³)	acid ⎫	1,430	320	(811) [e]
	7 (150 m³)	acid ⎭			(all tanks)
FRANCE					
La Hague		acid ⎫	1,400	133	(N/A)
		vitrified (glass) ⎭			
Marcoule				130	(N/A)
JAPAN					
Tokai				17.5	(N/A)
CHINA					
Jiuquan ⎫				7.6-15	(N/A)
Guangyuan ⎭					
GERMANY					
Karlsruhe				5.7	(N/A)
ISRAEL					
Dimona				2.4-4.2	(N/A)
BELGIUM					
Mol				4.1	(N/A)
INDIA					
Prefre				1.7	(N/A)
PAKISTAN					
New Labs			production (if any) unknown		

(a) Except for the U.S., values are estimated undecayed curies of Sr-90 and Cs-137 based on plutonium production and 3 curies each of Sr-90 and Cs-137 generated per gram of plutonium produced.

(b) An upper bound based on total capacity of all tanks (300 m³ x 100 tanks).

(c) Reported amount of all waste: 823 million curies in tanks plus 122 million curies of Sr-90 and Cs-137 in Lake Karachay and Staroye Marsh (Soyfer et al. 1992).

(d) Decay-corrected amounts as reported by the U.S. DOE 1991b.

(e) Total radioactivity as reported in U.K. Nirex 1988. Cs-137/Sr-90 estimated.

waste and the addition of other materials to the high-level waste tanks (the last being especially true of Hanford). The radioactivity per unit volume of high-level waste ranges from about 3,500 curies per cubic meter in the U.S. to somewhat more than 500,000 curies per cubic meter in Britain and France.

High-Level Waste Discharges into the Environment

LEAKS TO THE ENVIRONMENT

Although the early instances of discharging highly radioactive wastes directly into the ground were stopped, leaks of high-level wastes from tanks into the environment have been a problem.

In the U.S., despite early claims that the underground storage tanks at Hanford would be serviceable for "decades" and public assertions in the late 1950s that "none has ever leaked," tens of thousands of gallons of high-level waste had already leaked at the time such claims were made, and by the late 1980s an estimated 750,000 gallons containing significantly more than half a million curies had leaked out.[50]

In the U.K., also, there have been a number of problems with leaking tanks. A cladding silo sprang a leak in 1972, which was not detected until 1976. By 1980, it was leaking an estimated 185 gallons per day and had given out 50,000 curies of mostly strontium-90 and cesium-137.[51]

A leak of a HAST tank was discovered in 1979, amounting to 100,000 curies. The radiation field beneath the building was as high as 600 rads per hour, an extraordinarily intense field.[52]

WASTE DISCHARGES IN THE U.S.: HANFORD[53]

In the 1940s and 50s high-level wastes were dumped or injected into the ground at Hanford. The total volume of low-, intermediate-, and high-level liquid wastes disposed of in this way was 444 billion gallons or about 1.7 billion cubic meters. The total radioactivity of these wastes was 678,000 curies (not corrected for decay), of which 13,600 curies are from

---

50 Reviewed in Saleska et al. 1989, pp. V-1, V-2. U.S. General Accounting Office 1989 reports the U.S. DOE's estimate of 751,000 gallons, but does not cite radioactivity levels. Gillete 1973 reports that as of 1973, over half a million curies had leaked.

51 Carter 1987, p. 237.

52 Large 1992a; Carter 1987, p. 238.

53 Todd Martin, staff researcher, Hanford Education Action League, personal communication, June 30, 1992; U.S. DOE 1991c.

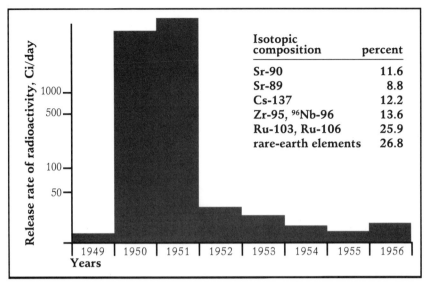

Figure 3.4. Discharge of radioactive wastes from Chelyabinsk-65 plutonium production to the Techa River (Chelyabinsk Region, Russia), in average curies per day, 1949–1956.

plutonium (184 kilograms of it), 40,500 curies from strontium-90 and yttrium-90, and 195,000 curies from cesium-137.

This dumping has contaminated groundwater in the area with radionuclides and other chemicals, and the contaminants are migrating, although there is considerable controversy about the exact levels of contamination and rates of movement.

WASTE DISCHARGES IN THE SOVIET UNION: TECHA RIVER AND LAKE KARACHAY

As indicated above, in the early years at the Chelyabinsk-65 plant, essentially *all* high-level wastes were discharged directly to the nearby Techa River.[54] This apparently occurred from 1948 through at least September 1951. After 1951, discharges to the Techa River drastically declined as plant managers began to discharge wastes into nearby lakes. However, significant amounts of "liquid radioactive wastes" apparently continued to be discharged into the river until at least 1956.[55] The total inventory discharged into the Techa River (1949–56) is reported by Soviet sources to be 2.75 to 3 million curies.[56]

54 Summarized from Soviet sources in Cochran and Norris 1992, pp. 32–33.

55 Kossenko et al. 1992; Soyfer et al. 1992.

56 Kossenko et al. 1992, Table 1; Soyfer et al. 1992.

According to Soviet reports, 99 percent of the activity was deposited in the first 35 kilometers downstream. But in 1951, elevated radioactivity ascribed to the Techa River contamination was found as far away as the Arctic Ocean — over 1,000 kilometers away.

In May of 1992 Saleska measured radioactivity at several points along the river and found elevated levels. For example, on the grassy banks of the river at Muslyumovo, a small farming village 78 kilometers downstream from the Chelyabinsk complex, his readings were consistently 300 to 500 microrads per hour, or 30 to 60 times natural background radiation. (This is an area where farm animals feed and children play; the village was never evacuated.) About 15 to 25 kilometers further downstream, on the silty riverbank where the main highway north out of Chelyabinsk crosses over the Techa, the readings were up to 8,000 microrads per hour, or about 500 times natural background radiation.[57]

Beginning in about 1952, most of the high-level waste radioactivity was instead dumped into Lake Karachay, which had no outlet, so the radioactivity was isolated from the Techa River system. This continued for at least a year before an intermediate waste storage facility was put into operation in 1953. Even after this date, however, the Soviets apparently continued to discharge medium-level wastes (i.e., the cesium-containing liquid in the tanks) into the lake. In the 1960s it was discovered that radioactivity from the lake was entering the groundwater.[58] The contaminated groundwater plume is beneath the radioactive waste burial grounds, and the disposal site is estimated to be 10 square kilometers in area. The volume of water is 4 million cubic meters (over 1 billion gallons), with a gross beta activity of 6,000 curies.[59]

The cumulative radioactive inventory of the lake reached 120 million curies, and the lake currently has an average surface radiation exposure level of 3 to 4 rads per hour. This jumps to 600 rads per hour near the discharge line — enough to provide a lethal dose in just one hour to anyone in the vicinity.[60]

57 Saleska 1992c.

58 Summarized from Soviet sources in Cochran and Norris 1992, p. 36.

59 Gorbachev Commission 1991.

60 Cochran and Norris 1992, p. 37. About 400 rems (4 sieverts) of dose delivered in a relatively short period will cause death within days or weeks to about half the people so exposed. This is known as the LD50 dose. We use the colloqial term "lethal" dose to mean anything on the order of greater than 400 rems delivered in a short period. Somewhat lower figures for LD50 are given by Joseph Rotblat 1992 (200 rads) and Fujita et al. 1989 (230–261 rads).

In 1967, there was a drought in the area, and evaporation exposed large areas of lakebed. A windstorm spread the dried sediment, containing about 600 curies of cesium-137 and strontium-90, over about 1,800 square kilometers, affecting some 41,000 people.[61]

There was also an explosion in a high-level waste tank in 1957 (see Chapter 4).

Together, these large discharges of radioactivity totaled 223 million curies. Of this, 5 million curies in all were dispersed over a wide area (26,000 square kilometers), affecting a population of about 450,000, according to a 1991 report for President Gorbachev.[62]

STUDYING HEALTH EFFECTS OF WASTE DISCHARGES FROM CHELYABINSK-65

In an unpublished report, one of the chief medical scientists at the Urals Radiation Medicine Research Center (formerly the Institute of Biophysics, Branch 4) in Chelyabinsk estimated that 124,000 people were exposed to elevated levels of radiation from living along the banks of the Techa River.[63] The river was the main source, and sometimes the only source, of household and drinking water for the inhabitants of the riverside villages. Residents fished, bathed, swam, and washed clothes in the river. River water was also used by cattle and domestic fowl.

Medical examinations of residents along the Techa River were reportedly conducted beginning in 1951, two years after radioactive waste dumping began.[64] Attempts at dose reconstruction centered initially on estimating levels of strontium-90, a bone-seeking element, using metabolic models; little biologically-relevant data on individual dose was available. Later, dose reconstruction was derived from measurements of surface beta-activity in teeth (begun in 1962) and then from whole-body counting (begun in 1974). Laboratory and outcome data that have been collected include studies of peripheral cell counts, chromosomal aberrations, birth outcomes, cancer mortality, and overall mortality.

A report edited by a member of the Supreme Soviet of the former Soviet Union claims that the "overall death rate of the people who lived along the upper Techa River was 17–23.5% higher than" in a similar group not

60 Cochran and Norris 1992, p. 37.

62 Gorbachev Commission 1991, Section 2.1.2.

63 Kossenko 1992a.

64 Kossenko 1992b.

exposed to radiation.[66] Nazarov and his colleagues assert that there are villages where the majority of people may have been affected by radiogenic diseases.[65]

In the last two years, as official secrecy has begun to lift, several radiation health specialists have come forward, presenting at international conferences and in international journals analyses of health conditions in and near nuclear weapons facilities. Nonetheless, formal epidemiological studies have yet to be published. At this stage, it is difficult to interpret the studies we have seen for several reasons: lack of any independent assessment of the quality and validity of the raw data, lack of detailed discussions of the methology used, and the usual problems encountered in environmental epidemiology (as discussed in Chapter 1).

The crack in Soviet/Russian secrecy regarding health and environmental effects of nuclear weapons production may afford the opportunity of collaborating further with Russian scientists and resolving or clarifying the difficulties of interpretation.

Unfortunately, we can expect efforts at obfuscation to develop as well. The U.S. DOE and Department of Defense (DOD) have shown great interest in the Chelyabinsk data on health effects of long-term radiation exposure. These two agencies have invited scientists from various Russian scientific and radiation health organizations and institutes (including the Urals Radiation Medicine Research Center) to the U.S. for consultation. Notable among these consultations was a workshop held in June 1992 at George Mason University in Fairfax, Virginia. The Russian and U.S. participants discussed their respective interests in a possible research collaboration and signed an agreement that could lead to a more formal collaborative arrangement or set of "exchanges." (It is noteworthy in this context that in 1991 responsibility for carrying out health studies around U.S. nuclear weapons plants was transferred from the DOE to the Department of Health and Human Services as a result of vigorous protests by physicians, scientists, and others about the poor quality of DOE's work in this area.)[67]

At first glance the objectives of such collaboration, as indicated in the workshop summary, appear to be purely scientific. For example:

> The Chelyabinsk data is expected to complement and refine the existing knowledge based on data collected in Japan following the Hiroshima and Nagasaki nuclear events, as well as data which were collected for Chernobyl

65 Nazarov et al. 1991, p. 49.
66 Nazarov et al. 1991, pp. 48–49.
67 Physicians for Social Responsibility 1992.

Figure 3.5. Inhabitants of the village of Muslyumovo, on the banks of the Techa River, looking at some of their first Western visitors, 1992. Muslyumovo is situated downstream of the Chelya-binsk-65 complex, which dumped high-level waste from plutonium production into the Techa from 1949 into the 1950s. Villagers were told not to drink the water but, given no reasons until 1989, they continued to use it for their daily needs. Other villages along the Techa were evacuated after health effects became obvious; Muslyumovo, although one of the closest to the point of waste discharge, was not. Photo by Robert Del Tredici.

and other accidents. It will be productive to compare the data from long-term studies of the Chelyabinsk region with data of more recent accidents such as Chernobyl. . . .

The focus of the meeting addressed reconstruction of dose and health effects based on more than forty (40) years of detailed medical and physical-chemical data collected at Chelyabinsk.

The U.S. delegation expressed strong interest in developing a collaborative scientific exchange to define health effects due to low exposures to radiation. . . .

It is anticipated that the data from Chelyabinsk will offer the U.S. a unique opportunity to develop a technical basis for the inception of radiation-induced health effects due to low levels of exposure. The Russian findings on the effects of low doses of radiation can help to reconsider risk-based standards.

Studies of the health effects of low-level radiation have been very diffi-cult and remain controversial for a number of reasons. Only one of these relates to the limited nature of the human database, derived principally from the people exposed suddenly during and shortly after the atomic bombings of Hiroshima and Nagasaki. Another fundamental reason is that data on workers in nuclear weapons plants are generally of poor quality, especially as regards internal burdens of and exposures to alpha- and beta-emitting radionuclides. Enormous uncertainties derive from the poor quality and incompleteness of the data. For example, a study of 96,000 nuclear workers in the U.K., based entirely on external exposures, indicates the cancer risk from low-level exposure was about the same as that yielded by Hiroshima and Nagasaki data. But the uncertainties were so high that a 90 percent confidence level ranged from less than zero risk to a risk almost two-and-a-half times that indicated by the Hiroshima and Nagasaki data. In the U.S., also, the quality of the data is seriously deficient. We are skeptical that the quality of the data in the former Soviet Union, where concern for people and workers around weapons plants was often far lower than in the U.S., will be good enough to resolve issues related to risk from low-level radiation.

The political interest in portraying Russian data as a convenient solution to a very difficult scientific problem was made clear by a DOE spokesperson in a radio interview with a journalist from National Public Radio, which was broadcast on June 17, 1992:

DAN CHARLES (reporter): All the official estimates of the risks of radiation are based on the experience of people who survived the atomic bombs in Hiroshima and Nagasaki. But many scientists think this method overstates the dangers. They believe, based on studies with animals, that a sudden blast of radiation, like at Hiroshima and Nagasaki, produces far more cancer than the same amount of radiation spread over a longer period of time, such as at Chelyabinsk.

If this is true, the radiation from America's own nuclear installations, such as the nuclear weapons plants operated by the Department of Energy, aren't [sic] as dangerous as now thought.

The DOE and the Pentagon were so interested in the Russian studies that they paid for the conference [at George Mason University]. DOE officials say that the Russian data may lay the scientific groundwork for looser standards for radiation releases. And Don Alexander of the DOE says that looser standards mean it will be cheaper to clean up America's nuclear weapons plants.

DON ALEXANDER (DOE): If our clean-up is required to go to the very significantly low levels that . . . we're currently being driven to, it could cost the United States a trillion dollars or more.

What we need to do is to establish very realistic standards that we can work towards. And we believe that that data only exists in one place on the planet, and that place is at Chelyabinsk.

In fact, there is currently no sound scientific basis for hoping that an analysis of the radiation doses received by people living in the Chelyabinsk region will result in a downward revision of radiation risks. Many people may have died without even being followed up or properly diagnosed. The Russian data need careful, independent evaluation as to their quality before they can serve as the basis for any conclusions about radiation risks.[68]

The history of radiation exposure along the Techa River and in other areas contaminated by the operations of the Chelyabinsk-65 complex has, tragically, provided another occasion to study the health effects of long-term radiation exposure, while obligating responsible agencies and individuals to see that the victims receive medical treatment as these studies proceed. It is collaborations between Russians and international physicians and scientists not allied with the nuclear establishment that are to be encouraged, for the sake of both science and the affected people. Certainly the DOE, whose health and environmental record in the U.S. is notoriously lacking, should not be the agency to lead such collaborations. Any collaborative research efforts on the part of the U.S. government should instead be carried out under the auspices of the Department of Health and Human Services.

68 Furthermore, Alexander's figure of "a trillion dollars" for the U.S. nuclear weapons cleanup program is baseless and alarmist. The current official cost estimates for cleanup are $150 to $200 billion.

Chapter 4
Tank Explosions: Kyshtym 1957

O N SEPTEMBER 29, 1957, at 4:20 p.m. local time, an explosion occurred at a storage facility for high-level nuclear waste at the Chelyabinsk-65 nuclear weapons complex in the Southern Urals.[1] At the time, the complex was still secret and not shown on any maps of the Soviet Union. Kyshtym was a nearby large town marked on the map. For this reason, and because the accident itself was kept secret for decades, it came to be known as the 1957 Kyshtym accident. Since the accident and the complex are now well-known, we will call this the 1957 Chelyabinsk accident.

As we have described, the Chelyabinsk-65 complex has been the major center for plutonium production for the nuclear weapons of the former Soviet Union. Reprocessing of irradiated reactor fuel is still carried on there, ostensibly for the main purpose of meeting the requirements of the breeder reactor program.[2]

It is worth describing the history of what it took to make the Soviet and the U.S. nuclear establishments take this accident seriously. This history epitomizes the cavalier attitudes to safety that have prevailed in the nuclear weapons business even in regard to the most serious demonstrated dangers. Indeed, there has been considerable suppression of information, and in effect a curious sort of collusion in this between the Soviet and U.S. weapons establishments, for fear that a concerned public would raise uncomfortable questions about the weapons complex or shut it down.

The Soviet government did not admit that an accident had occurred until June 1989, almost 32 years after the fact. However, as we will see, it did evac-

---

1 Nikipelov and Drozhko 1990.

2 It seems highly questionable whether that program can be implemented in view of the intense opposition to it since Chernobyl and the lack of resources after the economic and political collapse of the Soviet Union.

uate people from the region. The Soviet government also began a number of radiobiological studies in the region after the accident. The results of some of these studies were published in the scientific literature in the Soviet Union, though without revealing that they were associated with an accident. From the documentation that existed in the Soviet Union, a Soviet biologist, Zhores Medvedev, was able to piece together a coherent account of the explosion, which occurred in a tank containing highly radioactive wastes.[3]

Medvedev's work provoked a great deal of speculation and analysis in the West. When Medvedev began publishing his findings in Britain in 1976,[4] part of the nuclear establishment dismissed them. Medvedev related that "some nuclear experts, including the chairman of the United Kingdom Atomic Energy Authority, Sir John Hill, tried to dismiss my story as 'scientific fiction', 'rubbish' or a 'figment of my imagination'."[5]

The U.S. Central Intelligence Agency had long since discovered that there had been a radiation accident in the Urals. In 1959 a secret memo noted as follows:

> In the winter of 1957 an unspecified accident occurred at the Kasli (N 55°54', E 60°48') plant.
>
> All stores in Kamensk-Uralsky which sold milk, meat, and other foodstuffs were closed as a precaution against radiation exposure, and new supplies were brought in two days later by train and truck. The food was sold directly from the vehicles, and the resulting queues were reminiscent of the worst shortages during World War II. The people in Kamensk-Uralsky grew hysterical with fear, with an incidence of unknown mysterious diseases breaking out. A few leading citizens aroused public anger by wearing small radiation counters not available to everyone.[6]

The CIA soon discovered that there had been an accident at the Chelyabinsk-65 complex, near Kasli and Kyshtym. This was kept secret — even at a time when U.S. propaganda was strongly anti-Soviet — until a request filed under the Freedom of Information Act by a non-profit, public interest group, Critical Mass, forced some documents into public view in

---

3 Medvedev thought that this might have been due to a nuclear criticality. (See Medvedev 1976 and Medvedev 1977.) In fact, the explosion was a chemical one. However, we must point out that crucial data regarding the contents of the tanks, notably regarding plutonium content, are still secret, and it is not impossible that a criticality contributed to or even set off the explosion.

4 Medvedev 1976.

5 Medvedev 1977.

6 U.S. CIA 1959.

1977.[7] At that time it became clear that the information presented by CIA sources did support Medvedev's conclusion at least to some extent.[8] Medvedev himself has always been very sure that his deductions that an explosion of nuclear wastes had contaminated hundreds or thousands of square miles were correct and stated in 1977 that it widely known among experts in the Soviet Union for a long time. He himself had known about it since 1958, he stated.[9]

The Explosion[10]

A rectangular concrete structure at the Chelyabinsk-65 complex housed 20 stainless steel tanks containing highly radioactive wastes from plutonium production. The tanks were cooled by water flowing in the space between the tanks and the outer concrete containment. There were numerous problems with the storage tanks. One of them was the loss of information as to what was going on in the system, since the measuring instruments had deteriorated and "were in unsatisfactory condition."[11] Further, due to the high radiation fields and design defects in the layout, the instruments were for all practical purposes inaccessible and could not be maintained.

The tanks initially contained liquid radioactive wastes but gradually began to dry out due to the heat generated by the radioactivity. As the wastes began to dry, the tanks began to float in the cooling water around them. (The gases generated in the tanks were vented.) This put unanticipated stresses on the pipes leading to and from the tanks, since the pipes and their connections to the tanks were not designed to accommodate the stresses occasioned by floating tanks. This led to breaches in the pipes and contamination of cooling water.

This contamination was detected, and since it was considerable, the water had to be treated chemically. At this point a choice had to be made between safety and production, since the chemical processing capacity at the plant was not sufficient to both process cooling water and allow for maximum plutonium production. The choice was made to continue plutonium produc-

---

7 Wilson 1977. By February 1961, the CIA was also aware "as early as 1954 that the water of the Techa River . . . had become highly radioactive." (U.S. CIA 1961.)

8 Parker 1978, p. 6.

9 Medvedev 1977, p. 762.

10 This description of the accident is based on Nikipelov and Drozhko 1990.

11 Nikipelov and Drozhko 1990, p. 7.

tion and to cool the tanks intermittently instead of continuously. This did not provide sufficient cooling. According to Nikipelov, the First Deputy Minister of the Nuclear Industry in the USSR, and his colleague Drozhko:

> The efficiency of this system was not enough and, moreover, there was no monitoring of [the system], since the readings of the instruments were erroneous. An investigation made after the accident by a special commission showed that the most probable cause [of the accident] was the explosion of dry sodium nitrate and acetate salts that formed as a result of the evaporation of solutions in the tanks because of the self heating when cooling conditions were disrupted.[12]

The tank that exploded was 300 cubic meters in capacity. It contained 70 to 80 tons of highly radioactive waste with a total radioactivity of 20 million curies.[13] The total force of the explosion has been variously estimated at between 5 and 100 tons of TNT equivalent.[14]

The explosive materials were sodium nitrate plus sodium acetate. The particular mixture of wastes in the tanks resulted from a process that had been used in the early years of operation of the plutonium production plant. This process discharged wastes containing high concentrations of sodium nitrate (up to 100 grams per liter) and sodium acetate (up to 80 grams per liter). Dry mixtures of sodium nitrate and sodium acetate are known to decompose spontaneously and violently at elevated temperatures (about 380 degrees C). However these mixtures can be ignited with a spark or by friction at temperatures considerably lower than that. A brief nuclear criticality cannot be ruled out as an initiator of the explosion, though it is unlikely to have been a major source of the explosive power.

In analyzing the implications of the Medvedev papers, Frank Parker, a member of the Board on Radioactive Waste Management of the U.S. National Academy of Sciences, estimated that "if Soviet reprocessing technology was less sophisticated than ours" it may have left "as much as 3% of the plutonium-uranium in the wastes stream."[15] The official Soviet literature that we have seen does not discuss the sensitive issue of whether a criticality could develop in the wastes. In the United States, too, this possibility has been downplayed by officials. For example, in the case of the tanks at Savannah River, an official safety analysis initially

---

12 Nikipelov and Drozhko 1990, p. 7.

13 Nikipelov and Drozhko 1990, p. 5.

14 Falci 1990 (as cited in Cochran and Norris 1992, p. 40) gives a range of 5 to 10 tons of TNT, while Nikipelov 1989 cites a range of 80 to 100 tons of TNT.

15 Parker 1978, p. 9.

totally ruled out the possibility of an accidental criticality, but when this was challenged, a revised official analysis admitted that it was not entirely out of the question.[16] The issue remains to be dealt with properly and fully both in the U.S. and in the case of the Chelyabinsk-65 accident.

The Effects of the Accident

RADIOACTIVE CONTAMINATION

The tank was completely destroyed and two adjacent tanks damaged. All the waste was expelled from the tank in the explosion. Present official estimates are that 20 million curies of radioactivity were released in all. About 90 percent of the waste fell in the vicinity, going directly onto the soil surrounding the tank location. About 10 percent (2 million curies) was carried by the wind in a plume up to 1,000 meters high. Official figures show that the main constituent of the radioactivity was cerium-144 and its daughter product praseodymium-144. Cerium-144 has a half-life of 284 days and is estimated to have constituted about two-thirds of the total radioactivity. Zirconium-95 (half-life 65 days) and its daughter product niobium-95 are estimated to have contributed about one-fourth of the total. The most important long-lived component is stated to be strontium-90, a calcium-like, bone-seeking beta-emitter with a half-life of 29 years. Together with its daughter product, yttrium-90, it is estimated to have constituted about 5.4 percent (about 108,000 curies) of the total radioactivity in the fallout plume.[17]

Table 4.1 shows the radioactivity in the fallout plume at the time of the accident.

The fallout plume contaminated an elongated area towards the northeast whose footprint was 300 kilometers long, with a width of 3 to 50 kilometers. The total area contaminated with at least 0.1 curies of strontium-90 per square kilometer was 15,000 to 23,000 square kilometers. This is about twice the density of strontium-90 from global fallout from atmospheric nuclear weapons testing. About 1,000 square kilometers of this (100 kilometers by 10 kilometers) were contaminated at a level of 2 or more curies of strontium-90 per square kilometer.[18]

The official data on the radioactivity released in the tank explosion show only a small amount of cesium-137 relative to strontium-90. Nor-

16 Makhijani et al. 1986 and 1987.

17 Romanov and Voronnov 1990, p. 17.

18 Romanov and Voronnov 1990, p. 12.

Table 4.1. Amounts of radioactivity in the fallout plume from the Chelyabinsk-65 tank explosion

RADIONUCLIDE	HALF-LIFE	AMOUNT CURIES	PERCENTAGE
strontium-90 + yttrium-90	29.1 years	108,000	5.4
zirconium-95 + niobium-95	65 days	498,000	24.9
ruthenium-106 + rhodium-106	386 days	74,000	3.7
cesium-137	30 years	7,800	0.036
cerium-144 + praseodymium-144	284 days	1,320,000	66

Source: Romanov and Voronnov 1990, p. 17.
Notes:
1. Half-lives are from ICRP 1983. In rows that show two radionuclides, the half-life shown is that of the first, parent element. That of the second, daughter element, is far shorter in all cases.
2. Radionuclides not shown present in trace amounts include plutonium-239, europium-155, and strontium-89.

mally, these elements are both present in about the same order of magnitude in irradiated reactor fuel rods and in high-level waste from reprocessing. The discrepancy in the waste tank fallout is due to the fact that cesium-137 was selectively removed from some of the tanks and dumped into nearby Lake Karachay.[19] Consequently, in Lake Karachay, the discrepancy is reversed, and there is a disproportionately larger amount of cesium-137 (98 million curies) relative to strontium-90 (20 million curies).

According to L.A. Buldakov, a member of the USSR Academy of Medical Sciences and one of the officials in charge of the Institute of Biophysics at the time, cesium was extracted from nuclear waste for use as a gamma radiation source "for industrial and technological needs."[20] However, not all the cesium could have been extracted in this way, since there is a large amount in the reservoirs and Lake Karachay.

We might expect on the order of 1 kilogram of plutonium per year to have accumulated in the set of tanks such as the one that exploded, if the follow-

19 Presumably, most of the fission products (including strontium-90 but most likely not cesium-137) were allowed to precipitate and settle into sludge in the bottom of the tanks. The cesium-137 would have remained in solution in the liquid supernate, which was then apparently pumped off into the lake. This would have been similar to waste management practices at Hanford in the 1950s, when the liquid supernate at the top of waste tanks was pumped off and into the ground to make more room in the tanks. (However, in the Hanford case, the resultant environmental contamination was less severe, because the cesium-137 had first been precipitated out of the liquid and into the sludge at the bottom of the tanks via the addition of ferrocyanides.)

20 Buldakov 1989.

ing assumptions are true: 1) production at the plant was on the order of 1 ton per year (a reasonable estimate in view of total Soviet plutonium accumulation of over 100 tons in weapons), 2) extraction efficiency was about 99 percent, and 3) the tank had received all the wastes from plutonium production in that year. Thus, it is possible that several hundred curies of plutonium were present in the tank at the time of the explosion. (However, as mentioned above, Soviet plutonium separation may well have been far less efficient than 99 percent, leading to greater amounts of it in the wastes.)

Indeed, Bolsunovsky (1992) reports that the explosion deposited plutonium in the area at densities of up to 1 curie per square kilometer. This is not inconsistent with the above estimates.[21]

It is worth presenting relevant claims made earlier by the health official in charge as an illustration of the lengths to which the nuclear establishment would go to conceal health and environmental consequences of its activities. L.A. Buldakov claimed in a 1989 interview that there was essentially no plutonium in the fallout or in the soil around the exploded tank. He asserted that plutonium had been found in only one of a hundred soil samples at a typical concentration of as low as 0.01 becquerel per 1 to 10 kilograms of soil and that, therefore, it was difficult to tell whether this was due to routine operations or the accident. He further asserted that one should not expect to find plutonium because plutonium extraction efficiency at Chelyabinsk-65 had been 100 percent.[22]

Both statements are utterly implausible. In the first place, it is difficult to even detect radioactivity in concentrations as low as 0.01 becquerel per 1 to 10 kilograms.[23] Also, these concentrations are orders of magnitude lower than one would expect simply from fallout from atmospheric testing, even given the fact that some testing took place after the Chelyabinsk accident.

As for Buldakov's second assertion, a plutonium extraction efficiency of 100 percent is not possible in an industrial plutonium production operation even today. It was neither the case at Hanford nor at Savannah River. Given the more lax procedures in the Soviet Union and the lower attention to fine details of operation relative to the United States, it is unlikely that

21 Bolsunovsky 1992. By way of comparison, deposition of plutonium from nuclear testing fallout is 1.4 millicuries per square kilometer, or about one one-thousandth of this amount. (Eisenbud 1987, p. 335.)

22 Buldakov 1989.

23 For example, average natural background levels of radium-226 in soil are on the order of 40 becquerels per kilogram, which is 4,000 to 40,000 times Buldakov's claim of radioactivity deposited by plutonium from the explosion.

plutonium production efficiency consistently exceeded 99.5 percent. It is more likely that it was less, perhaps considerably less, efficient in the early, hectic years of the arms race. As mentioned above, Frank Parker speculated that as much as 3 percent of the plutonium-uranium may have been left in the waste stream from Soviet reprocessing. Further, we know that plutonium production methods were changed several times in those early years.[24] Given the lack of resources and the urgency with which plutonium production was being pursued, it is highly unlikely that production processes would have been changed unless efficiencies had been low. Recent official data indicate that one of these processes — the one that discharged the sodium-nitrate-sodium-acetate mixture into the tanks — had such an inefficient first-cycle extraction that a portion of the waste was sent back for "further extraction of uranium and plutonium."[25]

We have not calculated the health and environmental consequences of plutonium dispersal for the population exposed to the fallout. But in view of the recent findings that plutonium may be far more damaging than previously suspected, this is a matter that should receive careful scrutiny in future studies of the health effects.

DOSE ESTIMATES

Despite the very high levels of radioactive contamination in many areas, people were evacuated in phases over a period of about two years. About 270,000 people are estimated to have been in the fallout zone of about 15,000 square kilometers. In all, 10,180 people were evacuated over the course of these two years, according to official statements. Since the accident was kept a secret for over three decades, these people were not informed of the fate that had befallen them.

Official studies have now been published estimating doses to the exposed populations. Table 4.2 shows the doses to people who were evacuated, as related to contamination density and time of evacuation, according to official reports.[26] These official accounts to the IAEA did not estimate cancer fatalities due to the accident. Since it is a relatively straightforward exercise to make such estimates based on the official dose estimates, we have done so and shown them in the same table. We used the cancer risk coefficients from the

24 Nikipelov and Drozhko 1990, p. 6.
25 Nikipelov and Drozhko 1990, p. 6.
26 All the official reports we have seen contain much the same information. We have used Buldakov et al. 1989, Table 4.

Table 4.2. Dynamics of population evacuation and dose to the population before evacuation

| POPULATION GROUP AND SIZE | AVERAGE CONTAMINATION DENSITY, Ci/km² of Sr-90 | TIME TO EVACUATION, IN DAYS | AVERAGE DOSE RECEIVED UP TO EVACUATION, IN REMS | | POPULATION DOSE, EFFECTIVE DOSE EQUIVALENT IN THOUSANDS OF PERSON-REMS | EST. NO. OF FATAL CANCERS [b] |
			EXTERNAL EXPOSURE	EFFECTIVE DOSE EQUIVALENT		
1,150	500	7-10	17	52	60	47
280	65	250	14	44	12	9
2,000	18	250	3.9	12	24	19
4,200	8.9	330	1.9	5.6	24	19
3,100	3.3	670	.68	2.3	7	6
TOTAL: 10,730 [a]						100

Sources: Buldakov et al. 1989; values in last column calculated based on U.S. NAS 1990.
Notes:
(a) The estimated number of people evacuated as shown by this table taken from official publications is different from the total of 10,180 given in the text of the same source.
(b) Fatal cancers rounded to nearest whole number. We use a fatal cancer risk coefficient of 790 cases per million person-rem (10,000 person-sieverts) of dose. We use a dose rate reduction factor of 1 for solid tumors as the preferred way to estimate risk for these cancers. (U.S. NAS 1990, Table 4-2, risk factors for single exposures.) U.S. NAS 1990 says that slow doses may be less effective than sudden ones and are expected to reduce the risks of radiation exposure "possibly by a factor of two or more." The report also states, however, that at low doses the data are uncertain, and the risks could also be greater. The reduction of risks for low doses is based primarily on animal data, and recent data on British nuclear workers indicates that this hypothesis may not be justified. (Saleska 1992b.)

latest report of the Committee on the Biological Effects of Ionizing Radiation of the U.S. National Academy of Sciences (known as the BEIR V report).[27]

While considerable quantities of contaminated food were disposed of after the accident, agriculture was resumed in some contaminated areas starting in 1961. In all, 1,300 tons of grain, 240 tons of potatoes, 100,000 kilograms of meat, 67,000 kilograms of milk, and 61 tons of vegetables were disposed of because they were too contaminated. About 59,000 hectares of agricultural land were withdrawn from production, but 40,000 hectares were returned to cultivation by 1978, with the rest still deemed unsuitable "owing to high levels of contamination."[28]

In the zone that was the most heavily contaminated, doses were quite high. Buldakov and his colleagues estimate an average whole body equiva-

27 U.S. National Academy of Sciences 1990.
28 Nikipelov 1989, pp. 3–4.

lent dose of 52 rems (0.52 sievert), and 150 rems (1.5 sieverts) to the gas-trointestinal tract. Based on these figures, we estimate that in the most heavily irradiated zone, the number of excess cancers expected is about 47. This is a large excess of cancers in a population of only about 1,150 people who were the most heavily irradiated during the first seven to ten days. We calculate that we might expect to find about 100 excess cancers in the total evacuated population of 10,730, or an increase of about five percent in overall fatalities from cancer, if we assume that, of all fatalities, about 20 percent would be due to cancer in the absence of any radiation from human-made sources or other artificial environmental carcinogens.

These dose estimates include doses from gamma radiation, contaminated food, and contaminated water. They are all estimated doses based on models and assumptions about people's behavior and on the average conditions that might be expected given the measured contamination. We might expect there to have been substantial variations among people and in the actual extent of contamination of, for example, homes and clothing in the highly affected areas. Buldakov and his colleagues estimate that "[t]hese doses can be doubled in view of the non-uniformity of contamination density and the conditions in which exposure occurred."[29] Thus, specific segments of the population may have considerably higher cancer risk than shown by our calculations. This would put total expected fatal cancers in the range of 100 to 200. These cancer risk figures would have to be reduced by a factor of two (to a total of 50 to 100) if we used a dose-rate reduction factor of 2 for solid tumors, to take account of the slow delivery of doses.

The actual extent of the non-uniformities may have been considerably greater. Data from testing fallout and from the Chernobyl disaster show that local radiation deposition may vary by a factor of 10 or even 1,000.[30] Thus, it is possible that some people suffered very high and even lethal does of radiation in the first days after the accident. Although there is no confirmed evidence of this, there is some indirect and anecdotal evidence in this direction. One CIA report dating from 1976 states that "the chief evidence of the explosion was the tremendous number of casualties in the hospital at Chelyabinsk. Many of the casualties were suffering from the effects of radiation."[31]

29 Buldakov et al. 1989, p. 2.

30 IPPNW and IEER 1991, pp. 9–20; IAEA 1986; UNSCEAR 1988.

31 CIA report, cited in Parker 1978, p. 7. The CIA report did not date the explosion properly, putting it in 1956.

The official risk studies do not reflect the significant levels of cancer risk in the irradiated population indicated by our calculations. These studies typically conclude that, despite the high doses, the medical follow-up of the irradiated population found no excess cancers or other adverse health effects and that in the few cases where there were excesses, they were so small as to be statistically insignificant.[32] These early assertions may not hold up, however, as more information becomes available. For example, since many of the information controls were lifted in 1990, more data on the exposed population has been made public, along with the results of investigations conducted by scientists at the Chelyabinsk branch (No. 4) of the Institute of Biophysics. Although their analyses are not yet complete, a preliminary epidemiological study by these investigators reports finding an excess risk of leukemia in the population exposed to radioactivity released from Chelyabinsk-65.[33]

Buldakov and his colleagues claim that cancer incidence in the affected population in the area of the fallout was more related to large-scale industrial pollution by vast metallurgical processing plants located in the region:

> It was noted, for example, from the morbidity around Chelyabinsk, that there was no connection between enhanced morbidity and dose rate. On the other hand, a clear and complete correlation was found between morbidity and releases of SO_2 to the atmosphere. Although SO_2 is not itself a carcinogen, it is extremely useful as a gauge of chemical contamination. Actual data show that when there are no SO_2 releases, morbidity amounts to 225 cases per 100 000 individuals per year, whereas in situations where SO_2 is released in amounts of 50 000, 100 000 and 150 000 t[ons] per year, the morbidity figures rise to 250, 275 and 300 cases per 100 000 respectively. Accordingly, on the Chelyabinsk-65 map the cancer mortality figures correlate not with the radioactive contamination track but with the location of the metallurgical and chemical plants.[34]

It is now well established that pollution from industrial complexes in the Soviet Union was very severe, so the assertion above is not implausible on its face. (But it would have been more convincing had the authors found a correlation between cancer incidence and a carcinogenic industrial

---

32 See for instance Buldakov et al. 1989, and Nikipelov 1989.

33 Kossenko et al. 1992. The study population includes those exposed as a result of the 1957 waste tank explosion, as well as those who received exposures due to releases of radioactivity to the Techa River in the early 1950s.

34 Buldakov et al. 1989, p. 7.

pollutant, instead of between general mortality and sulphur dioxide.)[35] It may in fact be that the health damage from such plants overshadowed that from radioactivity. However, this does not mean that the radioactivity did not cause any damage, as is implied in the quoted statement.

Worker Exposures

Data on workers at Chelyabinsk-65 have presumably been collected for decades by researchers at the Institute of Biophysics, Branch 1. Some of these scientists have begun to report on some of the data at conferences. However, we have not as yet been able to get what we consider to be reliable data on exposure of workers who cleaned up the immense contamination at the site after the accident. Eighteen million curies of radioactivity were spilled there. Most of these radionuclides had half-lives in the range of several months to one year, but over five percent of the radioactivity was from long-lived strontium-90 (and yttrium-90), if we assume that the composition of the waste that was spilled into the soil was about the same as in the fallout cloud. Thus, we would expect that the radiation hazards in the vicinity of the site were immense in the years that followed the accident. At the Savannah River site, spills of a few hundred curies of cesium-137 have produced high radiation fields, creating problems for clean-up.[36]

Moreover, it is not clear what was done with the soil after it was scooped up. In a 1989 interview, Buldakov said that the soil was put in tanks.[37] In a 1991 interview, on the other hand, he stated that it was "bulldozed together and water was added to minimize dust."[38] Gennady Romanov, Director of the Environmental Research Station at Chelyabinsk-65, said in a 1992 interview that some of the contaminated soil was simply covered up with fresh dirt, and some of it was buried.[39]

35 We also note that the degree of correlation between morbidity and sulphur dioxide emissions is so perfect that it warrants skepticism. According to the reported figures, each increment of 50,000 tons of sulphur dioxide emissions per year is correlated with an increase in morbidity of *exactly* 25 cases. Considering the uncertainties inherent in defining the population that should be associated with sulphur dioxide releases, the very rough correlation between sulphur dioxide releases and other industrial pollutants, as well as other factors, such a perfect correlation is highly unlikely.

36 Makhijani et al. 1986, Tables 1 and 2 of Part II.

37 Buldakov 1989.

38 Buldakov 1991.

39 Saleska 1992c.

There have been anecdotal reports that large numbers of workers per-
ished in the clean-up, but we have not been able to find any confirmatory
evidence.[40] According to Monroe, Medvedev has estimated recently that
"tens of thousands of workers have been involved in the clean-up of the
Kyshtym explosion site — where 18 million curies have been deposited —
and the burial of irradiated forests, as well as later efforts to cover Lake
Karachay with soil and concrete."[41]

Against these reports are repeated official assurances that there have been
no fatalities as a result of the accident. For example, in the 1992 interview,
Romanov emphasized how carefully planned the clean-up had been, that the
workers who did it were professionals, and that no worker involved in the
clean-up received more than the 5 rem per year allowable limit. He denied
rumors that untrained soldiers had been used and had received high doses.[42]

The Relevance of Kyshtym to Other Locations

As mentioned, the government of the United States discovered soon after
the accident that there had been some kind of large radiation disaster in
the Urals. However, it chose not to publicize this issue, despite the
intense hostilities of the Cold War, presumably because questions would
have been raised about the prospect of similar explosions in the U.S.
nuclear weapons complex. Eventually such questions, among others, were
instrumental in stopping plutonium production at Hanford at the end of
the 1980s, just as the Chernobyl accident was instrumental in stopping
the operation of the N-reactor of similar design at Hanford.

It is interesting to look at the official responses of the U.S. nuclear
establishment since Medvedev first mentioned the accident in November
1976 in an article in *New Scientist*.[43] He had not intended to create a sensa-
tion. His article was not about the accident as such but about dissidence
among scientists in the Soviet Union. He mentioned the accident simply
as an illustration, assuming that scientists in the West already knew

40 More than one person in a position to know has made such assertions to staff and
   consultants of the IPPNW International Commission during the course of the
   preparation of this study. They have requested that the figures and specific assertions not be disclosed until the data can be verified.

41 Monroe 1991, p. 22.

42 Saleska 1992c.

43 Medvedev 1976.

about it.[44] They did not, since the CIA documents were still secret. But the nuclear establishments took note, and after first being dismissive, they began to wonder and make some analyses about what had happened.

In 1979, J.R. Trabalka and his colleagues from Oak Ridge National Laboratory along with Frank Parker of Vanderbilt University, published a paper in the U.S. journal *Nuclear Safety* calling for "an exhaustive critical review of the Soviet literature" due to the "growing importance of nuclear power as a world energy source."[45] By 1980 Trabalka and others had concluded that the "presence of an extensive environmental contamination zone in Chelyabinsk province of the Soviet Union, associated with an accident in the winter of 1957 to 1958 involving atmospheric release of fission wastes, appears to have been confirmed. . ."[46] The article further cited Professor Lev Tumerman, a former head of the Institute of Molecular Biology in Moscow, who had emigrated to Israel in 1972, that there had been hundreds of civilian casualties and that several thousand square miles had been contaminated. Finally the article concluded that "[t]he available evidence indicates that the most likely cause of the airborne contamination was the chemical explosion of high-level radioactive wastes associated with a Soviet military plutonium production site."[47] The failure of cooling systems in a high-level waste storage facility was recognized as a possible cause, and references to the scientific literature going back to 1956 were cited.

Yet, the reaction of scientists at Los Alamos analyzing the accident in 1982, well after these conclusions were published, was dismissive. In a paper entitled "An Analysis of the Alleged Kyshtym Disaster," two Los Alamos scientists, Diane M. Soran and Danny B. Stillman, found themselves in "complete disagreement" with the thesis that an accident involving nuclear waste had occurred.[48] They dismissed the thesis of the Oak Ridge team, saying that contamination of the Techa River could explain the evacuation of people and the razing of villages, and that this, together with other heavy routine radioactivity discharges from Chelyabinsk-65 and other problems, such as acid rain, could explain the observed phenomena.

44 Medvedev 1977.
45 Trabalka et al. 1979, p. 206.
46 Trabalka et al. 1980, p. 345.
47 Trabalka et al. 1980, p. 345.
48 Soran and Stillman 1982, p. 1.

Why were Soran and Stillman so vehement in dismissing the thesis of a nuclear waste accident? Consciously or not, they may have been acting to protect the United States' nuclear weapons complex, where the potential for explosions and their possibly serious consequences were denied until well into the 1980s. Frank Parker had already noted in his 1978 evaluation of the CIA documents and other materials that tank explosions were now "likely to be one of the major questions raised by intervenors in the future. We need to have answers. Could it happen here?"[49] While Parker himself recognized that such explosions were possible, the seriousness of the issue was only officially admitted publicly in the U.S. in 1989, in response to independent investigations and Congressional pressure.

A complete disclosure of the Chelyabinsk-65 explosion data, along with site-specific data from other centers where high-level waste tanks exist, would allow assessments for other locations regarding the force of possible explosions, the quantities of radionuclides which might be spread in fallout, the potential cancers in surrounding communities, and the health dangers to workers during such an accident and in on- and off-site clean-up. Lack of adequate data from the former Soviet Union about the explosion and from most countries about the exact nature of the contents of their tanks, has prevented us from engaging in such assessments here.

It is significant that only since 1989, have the dangers of tank explosions risen to the top of the agenda of the U.S. nuclear weapons complex. Internal documents show that the DOE had been aware of these problems prior to 1989, yet when independent analysts pointed out the threats, the DOE downplayed the risks as low.[50] The years 1989 and 1990 saw many public revelations about the dangers of high-level waste tanks in the U.S., notably at Hanford. It was also in 1989 that the Soviet government first admitted to the 1957 Chelyabinsk explosion. Intense public pressure has finally caused the DOE to call this issue its highest priority, specifically in relation to the Hanford tanks. We discuss these and other issues related to tank explosions in the next chapter.

---

49 Parker 1978, p. 8.

50 The first independent analyses to publicly point out that some high-level waste tanks in the U.S. posed serious health and environmental threats due to explosions were published in 1986 and 1987 (Makhijani et al. 1986 and 1987). At that time the spokespersons of the Department of Energy acknowledged that fires and explosions were possible, but denied that these were grave hazards. (Wald 1986 and 1987.)

Chapter 5
The Potential for Explosions and Fires in High-Level Waste Storage Tanks

F IRES AND EXPLOSIONS are perhaps the highest-consequence risks of high-level radioactive waste storage, as illustrated by the explosion at Chelyabinsk-65. Such explosions may occur due to build up of explosive gases, chemical events in the high-level radioactive wastes, a failure of cooling, or some combination of these. Over the years, there have been incidents of burns or explosions of high-level waste chemicals. Although these have not been catastrophic, they nonetheless provide a basis for concern that more serious accidents may occur.

Acidic and Neutralized Wastes

The factors affecting tank safety with respect to explosions vary widely. The first important distinction to be made is between acidic and alkaline wastes. When first exiting from the reprocessing plant, all high-level waste is strongly acidic, consisting largely of highly radioactive fission products in a nitric acid solution. However, their acidity makes raw reprocessing wastes more expensive to contain in the short term (for example, corrosion-resistant stainless steel tanks are needed, which are much more costly than simple carbon steel). Thus, sometimes wastes have been neutralized or made alkaline instead of being stored as an acid. All high-level waste storage presents some risks, but the nature of these risks is different depending on whether alkaline wastes or acidic wastes are being considered. Before discussing in greater detail the various possible mechanisms that might contribute to explosions or fires, we will briefly outline the characteristics of the alkaline and acidic waste forms.

In order to cut costs in the early years of the Cold War, the U.S. government built carbon steel tanks for the wastes, which were first made alkaline by adding sodium hydroxide.[1] This has had a number of consequences. First of all, as soon as a solution is no longer acidic, many of the substances which were dissolved in the acid precipitate out (settle to the bottom) as sludge. Thus, instead of a well-mixed liquid, the waste becomes a combination of liquid and sludge. This can lead to uneven distribution of material, resulting in hot spots (if radioactive materials build up in one area) and introducing the risk of criticality (if plutonium-containing particulates happen to concentrate in one area). It also makes it more difficult to determine the actual contents of the tanks, because samples are less representative of the whole than in the case of an evenly mixed liquid solution. Also, as will be discussed is greater detail in the following chapter on long-term waste management, alkaline wastes are more difficult to solidify into glass than are acidic wastes.

Acidic wastes tend to pose fewer problems since they can more easily be solidified for long-term management and the potential hazards introduced by making the waste alkaline are avoided. However, acidic wastes can also pose dangers. For example, since the stainless steel tanks necessary for storing acidic wastes are very expensive, there are strong economic incentives to minimize waste volumes. This means that the concentration of radioactivity tends to be much higher, and consequently, the wastes also generate much more heat. Continuous cooling of the waste is crucial.

The importance of the tank cooling system is illustrated by an accident at the French reprocessing plant at La Hague in 1980. La Hague experienced a plant-wide electrical failure in April 1980, when a fire and subsequent complications at the reprocessing plant knocked out both the regular and the emergency power supplies. Among the systems affected was of course the cooling system for the waste tanks, which contain radioactive wastes that are typically orders of magnitude more radioactive and therefore generate more heat than the average wastes stored in U.S. tanks. A cooling failure of three to ten hours could result in these wastes boiling, at which point they would begin releasing cesium-137 and ruthenium-

---

1   Often, the term "waste neutralization" is used. Technically, these wastes have generally not been made neutral but rather basic (or alkaline). A common measure of a solution's acidity or alkalinity is pH, on a scale of 0 to 14, with 7 being neutral. Solutions with a pH of less than 7 are acidic; solutions with a pH greater than 7 are basic (alkaline). Much of the waste stored in tanks at Hanford, U.S. has a pH of 10 or more.

106. These releases would contaminate the site and possibly the environment. The uncooled tanks could boil completely dry in a few days, possibly resulting in an explosion.

In this case, it was possible to bring an off-site generator to the site quickly, and the tank refrigeration systems were restarted after about one hour. Full power was not restored to the entire plant, however, for about ten hours. Even Cogema, the company that serves as the French nuclear materials directorate, referred to the fire and power failure as La Hague's gravest accident.[2]

A careful analysis of the chemical reactions or accident scenarios that might occur in the high-level waste tanks at sites around the world is necessary to adequately identify potential hazards associated with flammable or explosive chemicals that may be produced in these tanks. In order to predict the conditions that may result in rapid exothermic (heat-generating) reactions that can lead to explosions, it is necessary to know the constituents of the tanks. However, existing publicly available data is fragmentary, especially regarding countries other than the U.S.

In the U.S., also, information is limited (in many cases, it is not only that the information is kept secret, but that fully adequate data simply have not been collected and therefore do not exist anywhere). However, much more information has recently come to light there due to the ongoing extensive re-examination of the tank explosion risk that began in 1990. Since the information now available in the U.S. is probably the best available anywhere at this time, despite its limitations, we will use it a basis for our analysis. This will then provide a basis for future inquiries and questions that need to be raised for any place that high-level waste is stored.

Explosion Mechanisms

Only some types of high-level wastes stored in tanks can cause explosions. Some of the possible mechanisms for explosion in high-level waste storage containers are listed below. Poor cooling or a loss of cooling, such as happened at La Hague, increases the probability of some of these events.

• Explosions or burns resulting from the combustion of organic (carbon-containing) compounds in the tanks, which may have been introduced

---

2 *Le Matin* (April 18, 1980) and *Nucleonics Week* (April 24, 1980), as cited in Carter 1987, p. 317.

during reprocessing operations or added during waste management operations. In principle, the problem here is no different than in any other setting: organic compounds can act as fuel, which, in the presence of an oxidizer, can burn or explode if the conditions are right. Of particular concern in some tanks is the organic compound ferrocyanide.

- Hydrogen explosions, occurring from a build up of hydrogen and failure of ventilation in the tanks.
- Steam explosions or overpressurization events.
- Explosions or energy releases due to nuclear criticality (caused by the accumulation of a critical mass of plutonium in the waste).

We discuss each of these possible mechanisms in turn.

ORGANIC COMPOUNDS

The general risk with organic compounds is that they are a potential fuel and under the right conditions can burn or explode. In practice, the potential risks and consequences of a burn or explosion due to this mechanism vary widely and depend strongly on the specific chemical nature of the compounds involved, their concentrations, and the temperature.

Of particular concern at some tanks in the U.S. (and also in the former Soviet Union) is the presence of ferrocyanides ($Fe(CN)_6^{-4}$ compounds), which were added as part of waste management operations in the 1950s. In combination with nitrate and nitrite mixtures at high enough temperatures, ferrocyanide compounds can result in violent, potentially explosive exothermic reactions. We will first address the particular issues associated with the presence of ferrocyanides in tanks; this is followed by a discussion of issues associated with the presence of organics in general.

Ferrocyanides

Ferrocyanides were added to some single-shell tanks at Hanford, U.S. during experimental waste management activities in the 1950s, with the goal of concentrating wastes and saving tank space.[3] They were reportedly added to some storage tanks at Chelyabinsk-65, also.[4] The strategy was to precipitate much of the radioactive material (primarily cesium-137, as most of the strontium-90 and other fission products had already been made insoluble by neutralization of the acidic wastes) and dump the supernate (liquid part) into the ground or, in the case of Chelyabinsk-65, into waterways.

3 Program reviewed in Burger 1984, pp. 5–6, and Cash and Mellinger 1991.
4 Gorbachev Commission 1991.

If other organics are present, they may allow the exothermic reactions mentioned above to occur at lower temperatures and may themselves react exothermically with the nitrate or nitrite oxidants. DOE has reported results of laboratory experiments with "the most energetic mixture" of these various compounds showing a reaction threshold temperature of 220 degrees C, with explosion of unconfined milligram quantities occurring at a temperature of about 285 degrees C.[5] The position of DOE and its prime contractor at Hanford, Westinghouse, regarding ferrocyanide explosions has been that, although the chemical mixture in the tanks may be explosive, the tanks containing this mixture, at a maximum temperature of 57 degrees C, are not hot enough to induce explosion, at least under current conditions.[6]

There have been concerns expressed that the behavior of the large-scale system in the tanks cannot be reliably predicted by the laboratory experiments that have been conducted. This concern had been raised internally by DOE contractors at Hanford a number of years ago.[7] One 1983 report noted, for example, that "the large system size (a million gallon test tube) may cause the reaction to occur at a lower temperature than would be predicted based solely on observations in much smaller systems."[8] Although the authors recommended that more extensive follow-up studies be conducted to evaluate the potential for exothermic reactions involving ferrocyanides under various actual tank conditions, funding for such studies was not provided by DOE until the issue first came under strong public scrutiny and criticism in late 1989 and early 1990, shortly after the ferrocyanide explosion issue was finally made public for the first time in October 1989.[9]

A DOE advisory panel recently raised these concerns once again and criticized Hanford for placing too much reliance on temperature thresholds observed in laboratory experiments.[10] The panel reported that a number of factors were being neglected, noting in particular, "Each lab experiment which improves the realism leads to a reduction in the indicated minimum reaction temperature."

5 Gerton 1990.

6 Gerton 1990; Wodrich 1990.

7 Van Tuyl 1983; Burger 1984.

8 Van Tuyl 1983, p. 4.

9 The event that drew the public's attention was the release of a Hanford report (Burger 1984), which until that time had been kept secret.

10 Ahearne 1990, p. 5.

5.1. Aerial view of high-level radioactive waste storage tanks at the Savannah River Plant, Aiken, South Carolina, USA, 1983. Photo by Robert Del Tredici.

Another concern is that there may be "hot spots" in the tank — as a result of the highly non-homogeneous nature of the tank contents — that have not been detected by the single fixed vertical array of thermocouples DOE has thus far used to probe temperatures in each of the single-shell tanks.

Burger estimated that a worst-case accident in the ferrocyanide-containing tanks at Hanford could result in an explosion equivalent to 36 tons of TNT.[11]

Organic Chemicals

Ferrocyanides are not the only chemicals of concern in tanks. Organic chemicals in general provide a fuel that, at the right concentrations and temperature, can burn or explode.

The existing data on Hanford single-shell storage tanks is incomplete, especially in reference to the identity and amount of the organic constituents originating from the various primary recovery processes used over the last 40 years. Urgency for increased tank storage resulted in mixing of the stored waste, such that it has become very difficult to determine its chemical composition. Additional uncertainty arises from the magnitude of the waste transfer between tanks over the last 40 years.

11 Burger 1984.

The organic constituents in the waste resulted from five basic chemical processes: 1) the "Bismuth Phosphate Process," 2) tributyl phosphate solvent extraction of uranium from Bismuth Phosphate Process wastes, 3) the "Redox Process," 4) the "Purex Process," and 5) the B-plant fractionation process. Some wastes from other Hanford site facilities were also added to the single-shell waste tanks. The resultant waste is a complex mixture of a large number of organic complexants, organic solvents, and inorganic chemicals. Some of the products used in decontamination operations are proprietary, meaning that under patent laws their chemical make-up is secret. In summary, chemicals added in the above processes have not been satisfactorily audited.

The risk posed by chemicals in the tanks depends on a host of factors, many of which vary from tank to tank. One important factor, for example, is the pH of the waste, which sometimes determines what compounds can exist at all. For example, red oil (whose relationship to explosions during reprocessing was discussed in Chapter 3) is a metal nitrate in organic phase that cannot exist in alkaline solutions because of its chemistry. For this reason, it had not been considered an explosion hazard in the presumably alkaline wastes at Hanford. Recently, however, it has been found that some single-shell tanks at Hanford may have a much lower pH than had been generally thought.[12] Although the tanks were designed for holding wastes with a pH of 8 to 14 (and wastes in most tanks are generally maintained at a pH of 10 or higher), it appears that some tanks may have pH as low as 7[13] or even less;[14] there is as yet no adequate explanation of this phenomenon. Such conditions might allow for the formation of explosive red oils in tanks. Thus, the potential for problems similar to the red oil explosions discussed in Chapter 3 cannot be dismissed out of hand.

From the organic and inorganic chemical inventory, it is reasonable to assume that large quantities of organic material are intimately mixed with inorganic salts such as sodium nitrate ($NaNO_3$) and sodium nitrite

12 Wodrich 1991.

13 This issue came to light as a result of a recent review, by DOE contractors, of a 1952 study of the corrosion of steel by waste solutions in the range of pH 6 to 8. The document stated that some waste "is currently being stored at a pH of 7..." (Hanford Westinghouse 1952.) A subsequent review of historical sampling data taken since the mid-1950s from 107 of the 149 single-shell tanks at Hanford indicated that four tanks may contain wastes with pH in the range of 7 to 8.

14 John Tseng, Director, Hanford Program Office, Office of Waste Management, Department of Energy, personal communication with Arjun Makhijani, June 22, 1992.

($NaNO_2$). In addition, considerable modification of the original organic compounds may have occurred under the influence of temperature and radiation to produce partially oxidized compounds that are more readily susceptible to further oxidation (for example, hydroxylated and unsaturated organic compounds that would be particularly susceptible to reaction with nitrate ion).

At Savannah River, the accumulation of benzene in two tanks is an additional cause for concern. This benzene and other organics, including nitrobenzene and biphenyl, are released during the disintegration of sodium tetraphenylborate, which is used to precipitate cesium-137 prior to vitrification. (The vitrification plant is not yet in operation.) According to official estimates, a benzene explosion at Savannah River could release 8.6 million kilocalories, equivalent to about seven tons of TNT.[15]

HYDROGEN

Hydrogen is a combustible gas that is constantly being generated in virtually all high-level waste tanks by radiolysis of water and other hydrogen-containing compounds (such as organics, where they have been added). The degradation of organic complexants through radiolysis and chemical processes adds significantly to the generation of hydrogen and other gases such as nitrous oxide, nitrogen, and carbon dioxide.

Normally, the hydrogen and other gases bubble out of the waste and are removed from the air space in the tank by means of mechanical ventilation systems which filter out radioactive elements and discharge the hydrogen harmlessly to the atmosphere. However, in the event of ventilation system failure or other mechanism which holds up the release of hydrogen, flammable or explosive concentrations may result.

Hanford, USA

In March of 1990 DOE officials at Hanford publicly acknowledged that hydrogen in certain tanks presented a significant explosion risk;[16] documents show that contractor personnel at Hanford had been aware of the phenomenon of excess hydrogen build-up in at least one tank — tank 101-SY — since 1977.[17] The concern at Hanford is that in some tanks — 5 double-shell and 15 single-shell have been identified — mixtures of hydrogen and other gases tend to get trapped below the surface of the waste and cannot be safely

15 Du Pont 1988.

16 See, for example, New York Times 1990.

17 As acknowledged by Blush 1990, p. 4. Blush claimed that Hanford contractors had failed to inform DOE of the situation.

removed by the tank ventilation system. The hydrogen and other gases build up, raising the level of the waste surface, until they are released in a sudden "burp," after which the waste surface falls to its previous level.

Most of the analysis of this problem so far has focused on tank 101-SY, a 1.1-million-gallon double-walled tank believed to have the worst hydrogen accumulation problem. The gas build-up and release cycle for this tank is about 100 days, and as a result of the "burp," the hydrogen concentration in the tank exhaust vent approaches and even surpasses the lower flammability threshold for hydrogen of about 4 percent of volume (for example, over the last 1–1/2 years, "burps" have produced hydrogen levels in the tank exhaust system ranging from 0.5 to 4.7 percent). Furthermore, data from a May 1991 venting suggest that the maximum hydrogen concentration in the dome space of the tank may be greater than 5 times the level measured in the exhaust system. If reliable, this implies a maximum hydrogen concentration ranging from about 2.5 percent to about 24 percent, depending on the release event. The upper end of this range is not only above the flammability threshold of 4 percent but above the hydrogen detonation threshold concentration of 18.3 percent. At these times a single spark could initiate a hydrogen burn or explosion.[18]

Normally the pressure in waste tanks is kept slightly negative relative to atmospheric pressure, but when there is a burp, pressure can build up in the dome space. This can result in leakage of gases containing radioactive materials to the outside air from vents and other leakage points, potentially exposing workers to radiation or hazardous chemicals.

DOE and its contractors have said they do not know enough about the situation in the tanks to estimate the probability of a hydrogen explosion.[19] Nonetheless, DOE continues to assert that the probability of an explosion is "low." For example, one recent report stated, "The general consensus among technical reviewers of tank 101-SY is that the probability of an explosion is low. This is based primarily on the fact that combustion has not occurred during the last 13 years of operation."[20]

It is worth pointing out that according to rare event theory, if nothing else is known about the probability of an event other than the fact that it has not occurred in *t* years, the failure frequency can only be said, with 95

18 Saleska 1991.

19 U.S. DOE 1990a, 1990b.

20 U.S. DOE 1990b, p. iii.

percent confidence, to be less than 3/t.[21] Applying this to the situation at Hanford tank 101-SY, it is only possible to say (at the 95 percent confidence level) that the annual risk is less than roughly one in five. By most criteria, this would not be considered a low risk for an event of unknown but possibly catastrophic consequences.

Regarding the possible energy release from a hydrogen explosion at the Hanford tanks, we estimate from DOE reports of waste level fluctuations in tank 101-SY of up to one foot (in a tank with an inside diameter of 75 feet) that the gas build-up is on the order of 4,500 cubic feet of gas.[22] Assuming this is 50 percent hydrogen, the detonation energy would be approximately 1.5 x 10^5 kilocalories.[23] This is equivalent to the energy released in detonating 0.123 tons (roughly 270 pounds) of TNT. This could be taken as a worst-case scenario if ignition occurred at the time of greatest gas build-up and all hydrogen was consumed in the explosion, but, on the other hand, it does not include consideration of subsequent possibly explosive exothermic reactions involving organics and nitrates that might be initiated by the hydrogen explosion.

A recent re-analysis for DOE of the consequences of a hydrogen burn in one of the Hanford tanks concluded that, assuming certain unlikely conditions, there might be "possible structural failure."[24] In estimating the consequences under these conditions, however, the report assumed "that no significant structural damage occurred."[25] Even so, an estimated 2,230 kilograms of material is estimated to be ejected from the tank through the ventilation system network, most of it falling nearby, resulting in "significant ground contamination."[26] The maximum immediate dose received by the

21 As discussed in Du Pont 1988, p. 5-3. This figure is based on the assumption of a Poisson distribution.

22 Gerton 1990.

23 The density of H_2 of 0.0052 lb./ft.^{3} at 70° F is taken from Heung 1982, as cited in Du Pont 1988. The heat of combustion, 28,670 cal./g., is taken from Table 3-202, p. 3-142, of Perry 1963. The calculation is as follows: (4,500 ft.^{3})(0.5 H_2)(0.0052 lb. H_2/ft.^{3})(454 g./lb.)(28.67 kcal./g.) = 1.5 x 10^5 kcal.

24 Sullivan et al. 1992, p. 5, 6-5. The scenario envisioned leading to this was a hydrogen burn beginning under the crust covering the waste in the tank. It is based on the unlikely assumption that the crust is a solid, integral mass. In fact, based on sampling that has been conducted, it appears that the crust is more like a viscous fluid with suspended particles rather than a solid mass; thus, an under-the-crust burn is now generally thought to be implausible.

25 Sullivan et al. 1992, p. 7-5.

26 Sullivan et al. 1992, pp. 8-3, 8-4.

public in this scenario would be received by people on a nearby highway, and is estimated at 36 millirem.[27] Another accident scenario considered in the study estimated a smaller total release (60 kilograms), but larger consequent doses (190 millirems on the highway) due to the fact that all of the material released was in the form of smaller, more dispersible particles.[28]

The probability of an ignition occurring which might actually cause such an accident is not known, but the re-analysis report again makes the assertion that "the probability of either type of hydrogen-combustion accident is small based on past history."[29] As we discussed above, this is highly misleading — the short (about 15 years) history of tank 101-SY provides no basis for firmly concluding that the probability of a combustion accident is small.

Savannah River, USA

A 1977 environmental impact statement on Savannah River described the sequence of events required for a hydrogen explosion to occur in a waste tank:

A hydrogen explosion in a waste tank requires the successive failure of several equipment or procedural safeguards:

- Failure of tank ventilation system.
- Failure of pressure alarm to detect ventilation failure or failure of operating personnel to heed the warning.
- Spark ignition in tank after explosive gases have been generated in the tank.
- Failure of procedural safeguards (in routine check of blower operation, routine measurement of hydrogen composition in gas space of waste tank, etc.) to detect and correct ventilation failure.

Based on estimates of individual probabilities of these conditions, a hydrogen explosion is estimated to have a probability of approximately 1×10^{-3} [one in 1,000] per year.

The waste tank explosion postulated ... involves failure and collapse of the tank roof. It is estimated that one tank explosion in 10 would result in such

27 Sullivan et al. 1992, Table 8-IV, p. 8-5. Dose on Highway 240, 3.9 kilometers southeast of Hanford tank 101-SY.

28 This second scenario is one in which ignition begins in the dome space of the tank (above the crust) and some of the organics in the crust also burn. The crust combustion results in the generation of much smaller particles, and all of the material released is estimated to be in the form of particles of less than 10 microns. A much smaller amount is released, and the ground contamination is slight, but the doses are higher as a result of greater dispersion. (Sullivan et al. 1992, pp. 8-3 to 8-5.)

29 Sullivan et al. 1992, p. 2-4.

an extreme accident. The probability of the waste tank *explosion* postulated for Incident 6 is therefore about 10^4 [one in 10,000] per year.[30]

A later report reduced this estimate by about a factor of 2 to 5×10^{-9} per hour or about 4×10^{-5} per year.[31]

These probabilities are calculated by the "fault-tree" method, in which various modes of failure are postulated and probabilities are attached to each event in the sequence that contributes to each failure mode, as described in the above quote. Some of these probabilities are well-founded, while others, such as the estimate that only one in ten explosions would result in a total roof collapse, have no empirical basis.[32] As a result, one cannot have much confidence in the overall risk estimate attached to such explosions.

In fact, hydrogen concentration in the dome space has reportedly reached or exceeded the lower flammability limit of 4.1 percent by volume on two occasions. Furthermore, on at least 20 other occasions, hydrogen concentrations ranging between 5 and 100 percent of the lower flammability limit have been detected. In each case, the cause was inadequate ventilation due to equipment failures, power outages, or planned outages.[33]

Official estimates place the energy of a possible hydrogen explosion at Savannah River at 1.2 million kilocalories, equivalent to about one ton of TNT.[34]

STEAM UNDER PRESSURE

There have been a number of incidents at Hanford in which the tank pressure has risen, sometimes enough to cause venting to the atmosphere. Hanford personnel believe that the cause of these pressure increases (which occurred in seven tanks between 1953 and 1968) was superheated sludge at the bottom of a tank, which caused a steam bubble to form. The steam forces its way to the surface and in sufficient quantity can increase the pressure in the tank.

The worst of these was a "violent reaction" that took place in a million-gallon single-shell tank in January 1965.[35]

30 ERDA 1977.
31 Du Pont 1988, p. 5-51.
32 Makhijani et al. 1987; Du Pont 1988.
33 Du Pont 1988, p. 5-51.
34 Du Pont 1988, p. 5-52.
35 Womack 1977, attachment, p. 1. The tank is identified as 105-A.

The earth in the immediate vicinity of the tank was reported to have trembled, and a temporary lead cover on a riser on TK-103-A was dislodged, allowing steam to vent from this opening for about 30 minutes.[36] Several gallons of waste were ejected onto the ground from a line connected to the tank, and radiation dose rates of 400 rads per hour — which in one hour would amount to a lethal dose — were measured one foot from the spill.

This incident created an 80,000-gallon bulge with a 50-foot diameter in the tank bottom's steel liner, raising it 8 feet off the concrete foundation. The liner apparently ruptured, creating a "substantial crack in the tank floor."[37] Radioactive sludge leaked underneath the bulge, creating an additional hazard due to the accumulation of radiolytic hydrogen and oxygen in the underside of the bulge, between the liner and the foundation. Remediation was seen as difficult due to the fact that "[n]o access to the bulge is easily available, due to extreme radiation hazards, remote location of the bulge, and the extreme hazard of explosive release of radionuclide contamination."[38] Ultimately, wastes were transferred out of the tank, and in 1979 the tank was classified by DOE as "interim stabilized."

This accident is now believed to have been caused by water leaking into the space between the tank bottom's steel liner and the concrete foundation. This water is then thought to have been heated by the hot tank bottom to steam, which built up, lifting the liner eight feet off the bottom and then rupturing it. Besides causing ejection and leakage of highly radioactive liquids, the incident increased the pressure in the tank, and some radioactivity was released through the ventilation system.[39]

NUCLEAR CRITICALITY

A nuclear criticality occurs if a sufficient mass of fissile material (plutonium and uranium-235) accumulates in one place to cause a nuclear reaction. Since there is always some plutonium left in the waste because of extraction inefficiencies, this may present a danger in some waste tanks.

The danger of a nuclear criticality has generally been discounted or minimized in official discussions regarding U.S. tanks. Criticality requires the confluence of a number of conditions, including optimum geometry,

36 Beard et al. 1977.

37 Beard et al. 1977.

38 Beard et al. 1977.

39 Wodrich 1990.

5.2. Radiation Day at Chelyabinsk Ecology School. May 19 is Radiation Day at the Chelyabinsk Ecology School, sited on a dead river at the edge of the city's metallurgical district. Each year on this day students practice safety routines designed to protect them from a sudden release of radiation from the Chelyabinsk-65 complex 160 kilometers to the north. Chelyabinsk City, Russia, May 19, 1992. Photo by Robert Del Tredici.

minimum concentration of neutron absorbers, maximum concentration of neutron moderators, and a significant concentration of plutonium in the waste. According to analysis by officials at Hanford, the plutonium concentration in the waste must be at least 3 grams per liter. The highest concentration measured in tank sludge, however, is about 0.3 grams per liter, and the operating limit (maximum amount allowed) for plutonium in tank solids is 1 gram per liter.[40] However, it must be recognized that the solids in the Hanford tanks are not homogeneous and thus make for major sampling difficulties.

We have not undertaken a detailed assessment of the possibility of nuclear criticality in tanks. Where wastes are stored in acidic form, as for instance in France and Britain, the plutonium remains dissolved and uniformly dispersed in concentrations far too low to approach criticality. The potential for criticality is generally much greater in alkaline wastes (even though alkaline wastes in the U.S. are much more dilute) because under alkaline conditions the plutonium can settle to the bottom as solid partic-

40 Wodrich 1990.

ulates. This allows for the possibility that the solid particulates might then become concentrated in the tank sludge, where they might build up to critical concentrations in some spots.

Further, as mentioned previously, the fact that sludge can be non-uniform also means that it is difficult to determine with confidence the range of concentrations of plutonium that might exist in sludge. (One does not know how representative a sample taken from a non-uniform sludge might be.) Thus, unless a careful sampling program is undertaken, critical amounts of plutonium might conceivably build up and not be noticed.

The risk of nuclear criticality can be reduced by agitating the waste to keep it well-mixed. (We should note, however, that continuous agitation depends on a continuous supply of power, which is always subject to failure — power failures have happened at La Hague, Hanford, and Savannah River, at least.)

While the problem of potential explosions has been known for some time in official circles and discussed in reports, some of these reports were suppressed until recently, while in others the significance of the problem was downplayed.[41] However, that situation is gradually changing, with the most dramatic official acknowledgment of the potential seriousness of the situation being made in a letter report written by the Advisory Committee on Nuclear Facility Safety, chaired by John Ahearne. The letter report, dated July 23, 1990, stated that "the waste tanks are a serious problem" and said that the "possibility of an explosion . . . must be taken seriously. . ." The committee called the present situation "a prescription for potential disaster" and said that "a situation of this type at a nuclear reactor would lead to ordering a shutdown" — also noting, however, that "one cannot 'shut down' the tanks."[42]

Hence, at various locations around the world, workers and nearby residents will be stuck with risks of tank fires and explosions for periods ranging from years to decades, even if all reprocessing were to be stopped today.

41 Van Tuyl 1983; Burger 1984.
42 Ahearne 1990.

Chapter 6
Long-Term Management of High-Level Wastes

I N MANY COUNTRIES around the world, current plans for the long-term management of liquid high-level wastes left behind from plutonium separation involve the conversion of these wastes to a stable form. A common process anticipated (and in some cases practiced) for this is vitrification, in which the waste is converted into a glass form. Most plants use a borosilicate glass, which contains more boron than common glass and is similar in composition to the "Pyrex" glass used in kitchenware. It is conceived that these wastes would be disposed of in a repository below the surface of the earth.

In the U.S. and in Sweden, where reprocessing of spent fuel from commercial nuclear power plants is not practiced, this same type of repository is to be used for the disposal of the spent fuel wastes from commercial nuclear power reactors.

The permanent disposal of radioactive wastes has been a long-standing and controversial problem in much of the world. The scientific, technical, managerial, and political difficulties presented by a more-than-a-million-year disposal problem are varied and enormous. The first section of this chapter will discuss some of the generic difficulties associated with high-level radioactive waste disposal. The second section will then discuss some of the problems peculiar to the preparation of liquid reprocessing wastes for long term management and disposal.

General Background on High-Level Radioactive Waste Disposal

HISTORY

One of the major problems associated with radioactive waste is the fact that much of it will be radioactive for hundreds of thousands, if not millions of years. It will thus require isolation from the human environment for a period far longer than all of recorded history.

When the first high-level nuclear wastes were produced in the 1940s, in the course of the U.S. Manhattan Project to construct the first atomic bomb, they were stored in what were at the time considered to be "temporary" storage tanks. There was no plan for permanent disposal, and as far as the U.S. government was concerned, the exceptional and pressing circumstances of World War II relegated such long-term issues to a low priority.

Over ten years later, with the exigencies and shortages of that war long past, the U.S. government made a definitive commitment to commercial nuclear energy by licensing the first commercial reactor in 1957 — again with no waste disposal solution yet in sight. In remarks before the National Academy of Sciences conference in 1955, for example, A.E. Gorman of the U.S. Atomic Energy Commission's (AEC) reactor development division acknowledged that the attitude of the Commission had been to "sweep the problem under the rug."[1] And as Carroll Wilson, the first general manager of the AEC, acknowledged much later,

> Chemists and chemical engineers were not interested in dealing with waste. It was not glamorous; there were no careers; it was messy; nobody got brownie points for caring about nuclear waste. The Atomic Energy Commission neglected the problem... The central point is that there was no real interest or profit in dealing with the back end of the fuel cycle.[2]

Today, 50 years after the first artificial sustained fission reaction, and over 30 years after the first U.S. commercial reactor began operating, a great many studies have been done at much expense, but the subject is still controversial and there is still no demonstrated long-term solution to the more-than-a-million-year disposal problem presented by nuclear waste.

The disposal of highly radioactive waste deep below the earth's land surface in mined geological repositories was the first form of disposal seriously proposed (in the U.S. in 1957),[3] and is today perhaps the most widely assumed (though not yet technically demonstrated) form of disposal throughout the world.

However, there have been many other alternatives proposed and researched over the years. Among the options that have been considered

---

1 Carter 1987, p. 54.

2 Wilson 1979, p. 15.

3 U.S. National Academy of Sciences 1957.

are disposal by shooting into space or emplacement under the Antarctic ice cap. Another approach is to emplace wastes in ocean bottom sediments (sub-seabed disposal) at depths of a few tens to 100 meters in areas that appear to have long-term stability.[4]

THE NATURE OF THE HAZARD

Radioactive waste presents a range of hazards and problems which vary depending on the time frame under consideration. Of course, it is most hazardous soon after it is first generated, when there has been little time for the radioactivity to decay. As time passes, the radioactive hazard decreases, but significant hazard levels nonetheless persist for very long periods of time (from tens of thousands to millions of years).

Different radionuclides are important at different times. The short half-life fission products, like iodine-131 (half-life, 8 days), dominate the health threats early on. Other elements, like ruthenium-106 become relatively more important at intermediate times (on the order of 1 year). For intervals of time longer than this, three kinds of radioactive isotopes are important:

1. Moderately long-lived elements such as krypton-85 (half-life, 11 years), cesium-137 (half-life, 30 years), strontium-90 (half-life, 29 years), and plutonium-241 (half-life, 14 years). These elements constitute the bulk of the radioactivity from a few to a few hundred years after discharge from the reactor. (Plutonium-241 decays into other radioactive elements, called "daughter products," with much longer half-lives);

2. Very long-lived beta- and gamma-emitting elements, including carbon-14 and long-lived fission products like technetium-99, iodine-129, and cesium-135, which have half-lives of thousands to millions of years;

3. Long-lived alpha-emitting elements like radium-226 (half-life, 1,600 years) and transuranics like plutonium-239 (half-life, 24,000 years).

Figure 6.1 shows the radioactivity of high-level waste from reprocessing spent fuel as a function of time, from one hundred to one million years after discharge from the reprocessing plant. This figure shows the contribution from a number of different radionuclides individually, as well as the total radioactivity. As can be seen, the contribution from some radionuclides (such as plutonium-239, uranium-233, radium-225 and radium-226) actually increases for a while before peaking and then declining.

4 A number of these alternatives are discussed in U.S. DOE 1979. For an overview of sub-seabed disposal see U.S. Congress OTA 1986 or Hollister et al. 1981.

This is due to the fact that these are being produced by the decay of other radionuclides.[5]

The extreme longevity of some of the radionuclides in high-level waste means that it is impossible to guarantee that this waste will remain completely isolated from the environment. "Isolation" then becomes a relative term in which it is assumed that some radioactivity will be released into the environment over time. In fact, according to the U.S. National Academy of Sciences, "[e]ssentially all of the iodine-129 [half-life: 15.7 million years] in the unreprocessed spent fuel in wet-rock repositories will eventually reach the biosphere."[6] And as the U.S. EPA has remarked, any environmental standards regulating "acceptable" releases of radioactivity from nuclear waste repositories must therefore "address a time frame without precedent in environmental regulations."[7]

Standards for radioactive waste disposal are thus usually based on the assumption that some radioactivity from a repository *will* reach the human environment.

CURRENT STATUS BY COUNTRY

In 1982, U.S. law established the framework for an elaborate approach to selecting two sites (in two different regions of the country) for the disposal of high-level wastes (both spent fuel and high-level reprocessing wastes) in a deep underground repository. These two final sites were to be winnowed from a larger number in a two-stage selection process in which objective technical grounds were to be used as the principal basis for discriminating between sites. This process was largely abandoned in practice, and in 1987 amendments to the 1982 law designated one site — Yucca Mountain, in the state of Nevada — to be characterized in depth as a potential high-level waste repository. Although the 1982 waste law mandated a target date of 1998 for the availability of a permanent repository, the U.S. DOE has since delayed the program twice. The current timetable

---

5 For example, americium-243 (half-life, 7,950 years) in the waste decays to plutonium-239 via alpha and beta decay. Since americium-243 has a shorter half-life than plutonium-239, it starts out producing new plutonium-239 at a faster rate than the already existing plutonium-239 is being eliminated by its own decay. The net result is that between about 1,000 and 20,000 years after discharge, the amount of plutonium-239 increases by perhaps a factor of two before the supply of americium-243 is exhausted and an irreversible decay takes over.

6 U.S. National Academy of Sciences 1983, p. 11.

7 U.S. Environmental Protection Agency 1985, p. 38066.

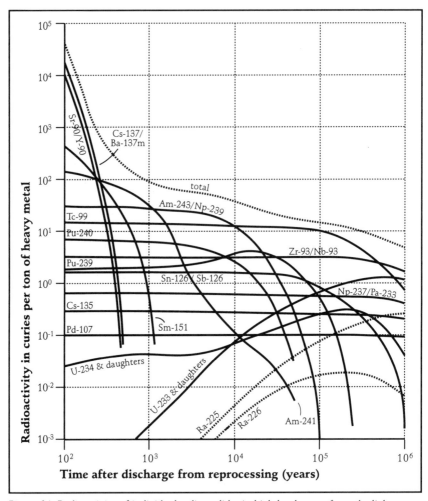

Figure 6.1. Radioactivity of individual radionuclides in high-level waste from the light-water reactor uranium fuel cycle. (Reprocessing, 150 days after discharge from reactor; enrichment, 3% uranium-235; burn-up, 30,000 megawatt-days per ton of heavy metal; residence time, 1,100 days; 0.5% uranium and 0.5% plutonium remaining in the high-level waste.) (Adapted from Benedict et al. 1981.)

is that, if the Yucca Mountain site is found suitable, it should be constructed and available to begin loading waste in the year 2010.[8]

Most other nuclear-power and nuclear-weapons states have also elected to put high-level radioactive waste into deep geologic repositories. Table 6.1 indicates the status of high-level waste programs in a number of coun-

8 U.S. DOE 1988, p. 1; U.S. DOE 1989a, p. vii.

Table 6.1. Programs for high-level waste burial in selected countries

COUNTRY	EARLIEST PLANNED YEAR	STATUS OF PROGRAM
Argentina	2040	Granite site at Gastre, Chubut, selected.
Belgium	2020	Underground laboratory in clay at Mol.
Canada	2020	Independent commission conducting four-year study of government plan to bury irradiated fuel in granite at yet-to-be-identified site.
China	none announced	Irradiated fuel to be reprocessed; Gobi desert sites under investigation.
Finland	2020	Field studies being conducted; final site selection due in 2000.
France	2010	Three sites to be selected and studied; final site not to be selected until 2006.
Germany	2008	Gorleben salt dome sole site to be studied.
India	2010	Irradiated fuel to be reprocessed, waste stored for twenty years, then buried in yet-to-be-identified granite site.
Italy	2040	Irradiated fuel to be reprocessed and waste stored for 50-60 years before burial in clay or granite.
Japan	2020	Limited site studies. Cooperative program with China to build underground research facility.
Netherlands	2040	Interim storage of reprocessing waste for 50-100 years before eventual burial, possibly sub-seabed or in another country.
Soviet Union	none announced	Eight sites being studied for deep geologic disposal.
Spain	2020	Burial in unidentified clay, granite, or salt formation.
Sweden	2020	Granite site to be selected in 1997; evaluation studies underway at Aspo site near Oskarshamn nuclear complex.
Switzerland	2020	Burial in granite or sedimentary formation at yet-to-be-identified site.
United States	2010	Yucca Mountain, Nevada, site to be studied and, if approved, to receive 70,000 tons of waste.
United Kingdom	2030	Fifty-year storage approved in 1982; exploration of options including sub-seabed burial.

Source: Lenssen 1991, pp. 24–25.

tries, including target dates for repository operation. As in the United States, public opposition to these plans has usually been intense. For instance, in France, where reliance on nuclear power is the greatest of any country in the world, the announcement of prospective sites set off such a furor, especially in farming regions that produce famous French gourmet foods, that the siting process had to be suspended.

Sweden does have an underground repository for low- and intermediate-level wastes from nuclear power plants but as yet no repository for spent fuel. The opposition in Sweden may be less intense than that in other countries in part because the program to dispose of nuclear waste is accompanied by a phase-out plan for nuclear power.

In the United States, there is another category of wastes that is supposed to be put in a geologic repository. This is the category of transuranic waste (referred to in some countries as plutonium-contaminated materials). The repository built by the U.S. DOE to take these wastes is called the Waste Isolation Pilot Plant. Unlike DOE's program at Yucca Mountain, WIPP is partly built.[9] Located 650 meters below the surface, the $1 billion repository consists of a 112-acre underground area, and has a capacity of about 880,000 55-gallon drums, enough to contain slightly less than 160,000 cubic meters (5.6 million cubic feet) of waste.[10]

Numerous technical issues related to the geology and hydrology of the WIPP site and the nature of the transuranic waste intended to be placed there raise questions about its suitability and about DOE's management of the program. These issues include leakage of water into the WIPP site, and a rate of waste room closure two to three times faster than anticipated. This has led to extensive wall cracking and numerous instances of ceiling collapses.

9 Only about 15 percent of WIPP has actually been mined. This is because the natural phenomenon of "salt creep" (which is the tendency of salt to gradually "flow" and fill empty spaces) causes any rooms mined to close as the salt creeps in to refill the mined space. This gradual room closure is an anticipated part of any waste-disposal process in salt, but the rooms cannot be mined too far in advance of waste emplacement. The DOE therefore plans to mine the additional waste-disposal rooms as the time of permanent waste emplacement approaches. (U.S. DOE 1989b, p. 2-7.)

10 The actual capacity of the repository has been the subject of some controversy. At one point DOE claimed that the capacity was 1.1 million 55-gallon drums containing about 6.5 million cubic feet (U.S. DOE 1989b). The New Mexico Environmental Evaluation Group, however, estimated that the space only allowed for about 850,000 drums, after which DOE apparently adjusted its estimate to 880,000 drums containing 5,598,000 million cubic feet (about 158,500 cubic meters). (U.S. DOE 1990c, Comments and response section of Vol. 3 and Table 3.1 in Vol. 1.)

The DOE recently tried to begin loading waste into the WIPP repository by short-circuiting the legal process for authorizing such steps through the U.S. legislature. So far, however, the courts have prevented it from doing this. As a result, the DOE is now attempting to obtain the necessary legislative authorization, even as the rooms of the salt mine that were supposed to contain the waste begin to cave in.[11]

Problems Associated with High-Level Liquid Reprocessing Wastes

There are a number of problems involved in the long-term management of reprocessing wastes that go beyond the general ones of long-lived radioactive waste disposal. We will discuss some specific potential difficulties associated with each of the following:

- conversion of high-level liquid reprocessing wastes into a stable form suitable for long-term management and disposal (e.g., the conversion of wastes via vitrification into a borosilicate glass form)
- the compatibility of the selected long-term waste form with the long-term disposal option eventually employed (e.g., the compatibility of borosilicate glass with the rock type of a deep underground repository)
- the removal of wastes from the underground storage tanks.

CONVERSION TO A STABLE FORM

The properties of glass, such as solubility, maintenance of integrity at high temperatures, and other aspects of durability, depend on its overall composition, including its boron content. Borosilicate glass contains high levels of boron (several percent) and is similar to the familiar "Pyrex" glass from which kitchenware is made. It has long been considered as a possible medium in which to mix highly radioactive wastes from plutonium production (civilian or military), in order to immobilize them for disposal in a geologic repository.

Vitrification (the formation of wastes into a borosilicate glass form) is planned in the U.S.; a vitrification plant has been built at Savannah River, and one is planned for Hanford and West Valley. A new pilot-scale vitrification facility has operated in the Soviet Union, and one began operating at Sellafield in the U.K. in 1992. Vitrification of acidic wastes has been practiced in France for a number of years. The French experience has led many to the conclusion that vitrification is a demonstrated, proven tech-

11 See Makhijani and Saleska 1992, pp. 51–58, for discussion of these issues.

Figure 6.2. Vitrification experiment, Savannah River Plant, Aiken, South Carolina, USA, 1988. Steel cylinders holding vitrified material have been sawed open here. A vitrification plant has been built at Savannah River, but its opening has been delayed at least until 1993. Photo by Barbara Norfleet.

nology. While the French plants have been operating for some time, we have no definitive data regarding any problems that may or may not have been encountered. In any case, as discussed further below, the vitrification of alkaline wastes — particularly at the Savannah River and Hanford plants in the U.S. — is indeed problematic.

Problems with the Vitrification Process

The chemical and physical form of high-level waste varies. Vitrification is not a proven technology with respect to all wastes. Depending on management practices, including chemicals added or removed, the degree to which different waste forms have been indiscriminately intermixed, and other factors, vitrification of high-level waste can be an extremely difficult and problematic proposition.

For example, at Hanford in the U.S., several different plutonium recovery processes and tank management practices have been used over time.[12]

12 Separation processes used at Hanford include the bismuth-phosphate process, the reduction-oxidation (Redox) process, and the plutonium-uranium extraction (Purex) process. The earlier processes produced far greater volumes of waste with somewhat different chemical composition. The early bismuth-phosphate process was not able to recover uranium. (Cochran et al. 1987b, pp. 14–15.)

Waste volume and composition have varied widely, partly as a function of the plutonium recovery processes used and partly as a function of waste treatment processes employed subsequent to plutonium recovery. As one Hanford official recently said about the Hanford waste-storage tanks, "No two are alike. You can never make a general statement about Hanford tanks, because somewhere you will be proven wrong."[13]

Waste treatment methods used at Hanford included:

- the addition of sodium hydroxide (NaOH) to neutralize the acidity of the waste, and allow storage in carbon steel tanks (which would otherwise be quickly corroded)
- the addition of ferrocyanides and phosphates to precipitate cesium-137 and strontium-90, respectively (i.e., to make them settle out as cesium ferrocyanide and strontium phosphate), to reduce the activity and allow the disposal of the tanks' supernatant liquids as low-level waste
- removal of strontium-90 and cesium-137 to reduce the radioactivity in the tanks and allow the remaining waste to be concentrated
- solidification of tank contents by evaporation of water to crystallize the waste into a salt cake form (this was done in some instances to prevent the loss of material to the environment from tanks that were leaking).[14]

Problems due to treatment methods and the differences among them have been the cause of a substantial delays in the U.S. vitrification program. A vitrification plant — called the Defense Waste Processing Facility — has been built at Savannah River, and although DOE originally planned to begin converting reprocessing wastes to radioactive glass in 1990, problems have caused delays.[15] DOE currently expects to open the Savannah River vitrification plant in 1993, although further delays may be possible.

The program at Hanford has encountered especially substantial problems. In some cases this is due to the explosive nature of the contents of some of the tanks (as with the tanks containing ferrocyanides). This may prevent solidification by conventional means. For example, heating up some of the waste mixtures (which is necessary for their vitrification) might cause explosion. Although much smaller quantities than a tank-full would be involved in the vitrification plant process stream at a given time, violent reactions could destroy or damage vitrification equipment and possibly lead

13 Remarks of Don Wodrich, Hanford Technical Exchange Program (November 13, 1991). (Saleska 1991.)

14 U.S. DOE 1987, p. 3.4.

15 U.S. DOE 1984, p. 64.

to release of radioactivity. At the very least, this poses severe problems and constraints for the design of the vitrification plant, as a recent DOE technical assessment of the Hanford vitrification program has concluded:

> As a result of a multiplicity of concurrent technical issues and disruptions... the detailed design of the [Hanford Vitrification Plant] is considered premature.... A re-evaluation of the programmatic objectives, technology basis, management philosophy, organizational structure, and cost is required before major actions can be prudently taken.[16]

Problems with Vitrification By-Products

The impression often left in discussions of vitrification is that all of the waste is simply converted to a form for disposal in a repository meant for high-level waste. In actuality, however, this is not necessarily the case. Prior to vitrification, sometimes substantial quantities of radionuclides are removed and disposed of as so-called "low-level" wastes. This is especially so in the U.S., where the volumes of high-level waste in the tanks are very large for reasons discussed in Chapter 3.

The U.S. DOE has planned to separate some of the radioactivity in Hanford's high-level liquid reprocessing wastes and convert it into a cement-like mixture called "grout" prior to vitrification. The U.S. DOE has been attempting to classify the grout as "low-level" waste, so that it can be buried in shallow vaults.[17]

The quantities of radionuclides planned to be turned into grout at Hanford as of early 1989 — although deemed "low-level" — were enormous in quantity and included:

- 12 to 20 million curies of cesium-137
- 1 to 8 million curies of strontium-90
- 30 to 150 kilograms of plutonium[18]

(As a point of comparison, according to current DOE records, the cumulative total radioactive inventory of all radionuclides in all AEC/DOE low-level radioactive waste disposed at all active DOE disposal sites in the U.S.

---

16 U.S. DOE 1991a, p. I-2.

17 The waste category into which such grout falls is a problematic issue. U.S. nuclear waste regulations (at 10 CFR 60) define high-level waste as including "liquid wastes resulting from... reprocessing irradiated reactor fuel," including "solids into which such liquid wastes have been converted." Thus, despite DOE's desire to treat it as "low-level" waste, the grout derived from liquid reprocessing wastes would seem to be categorized as high-level waste according to U.S. regulations.

18 Wodrich 1989.

through 1990 amounts to about 13 million decay-corrected curies.[19] And the entire grout plan at Savannah River envisages a grout discharge whose accumulation peaks at about 116,000 curies per year.)[20]

This is a huge amount of radioactivity that by any reasonable standard should be considered long-lived waste destined for a permanent repository. That such a proposal could be seriously considered shows again the danger in the present lack of standards in the U.S., and the need for ones which are clear and identifiable.

Recent experiments with Hanford grout are showing the potential danger of this approach. Organic chemicals added to the high-level waste as part of past waste treatment operations at Hanford are causing problems in the formation of the grout. Organic chemical breakdown is leading to the evolution of hydrogen gas from the grout, and Hanford personnel have had to put pipes into the experimental grout mixtures to vent the hydrogen gas. Thus, it seems likely the grouted waste form will be susceptible to rapid cracking and disintegration. Further, the experiments show that the organic chemicals and nitrates in particular are highly likely to leach out of the grout, posing a groundwater pollution problem, especially from the nitrates.[21]

We do not have comparable data for other sites.

COMPATIBILITY OF WASTE FORM WITH DISPOSAL METHOD

The geologic barrier plays a central role in the conceptual design of virtually any repository system designed to prevent releases of large quantities of radioactive materials to the environment. Geologic sites are typically highly complex and non-uniform. Each type of rock and geological setting possesses its own unique physical, chemical, and hydrogeological characteristics. These characteristics interact with the chemical forms of the

19 U.S. DOE 1991b, Table 4.1, p. 109. The sites (as listed in U.S. DOE 1991b, p. 115) where low-level defense wastes are buried include: Hanford, Savannah River, Idaho National Engineering Laboratory, Oak Ridge, Fernald, Nevada Test Site, Los Alamos, Lawrence Livermore, Paducah, Portsmouth, Sandia National Labs, and Brookhaven National Labs.

20 The peak of 116,000 curies is projected to occur in 2007 (U.S. DOE 1991b, p. 121). (Grout at Savannah River is generally referred to as "saltstone.")

21 This issue was discussed at the meeting of DOE's Technical Advisory Panel on Hanford high-level waste tanks, in Chicago on September 5, 1991. (Arjun Makhijani, personal notes.)

radioactive materials that would be released from the waste package. Therefore, the ability of a repository to contain wastes is connected to the nature of the waste form and waste package and their potential chemical and physical interactions with the rock and water in the repository.

For this reason, it is prudent to give detailed consideration to the inter-actions of waste form with potential repository settings before selection of the waste form. However, in many cases, glass has been selected as the waste form without any regard whatsoever for the location and nature of the repository host rock. Many countries have only barely begun to con-sider where to locate repositories for their waste.

It might also be considered prudent to select waste forms whose perfor-mance can be guaranteed to be very good under a wide variety of geo-chemical and hydrogeological conditions. This however, may not be the case with glass. As an example, we consider the U.S. situation with respect to the selection of the Yucca Mountain site in Nevada. Even with the incomplete and inadequate information now at hand, there are seri-ous questions as to the performance of glass at the Yucca Mountain site due to a phenomenon known as "hydration aging."

Hydration Aging of Glass at a Yucca Mountain Repository[22]

A 1981 DOE-appointed panel ranked borosilicate glass first among candi-date waste forms, whereupon DOE selected glass as the waste form for high-level waste at Savannah River. However, according to a 1983 report by the National Academy of Sciences' Waste Isolation Systems Panel, the criteria used by the DOE panel to rank waste form did not relate system-atically to waste form performance under repository conditions:

> It is premature to select waste form materials on the basis of such rankings. The effects of higher temperatures and the effects of realistic repository condi-tions could alter the rankings. Further, a number of alternative waste form materials have had little study. . . . [F]or most of the important long-lived radionuclides the laboratory leach data that have been used in these rankings have little relevance to the releases of radionuclides in a geologic repository.[23]

Work published in 1982 by J.K. Bates et al. raised just such a possibili-ty.[24] This work investigated "hydration aging" of glass under conditions that might prevail at Yucca Mountain. Hydration aging is a phenomenon

---

22 This issue is discussed in greater detail in Makhijani 1991, from which this section is adapted.

23 U.S. National Academy of Sciences 1983, p. 51.

24 Bates et al. 1982

that has been observed to occur on hot glass surfaces in the presence of steam, wherein the surface layers of glass disintegrate much more rapidly than otherwise. Plutonium releases under conditions of hydration aging could be several hundred times greater than without such aging. Surface layer disintegration is observable within a matter of weeks.[25]

The environment at Yucca Mountain is particularly susceptible to this problem if water enters the repository, because the repository would be at atmospheric pressure.[26] This means that near 100 degrees C most or all water would be in the vapor phase. This could result in hydration aging and potentially rapid disintegration of glass in violation of U.S. NRC regulations requiring some integrity of waste form for up to 100,000 years. The dose implications of such releases would range from small to enormous, depending on hydrogeological conditions and the geochemical reactions of the disintegrated material with the rock.[27]

Independent investigators have repeatedly pointed out the significance of this issue since 1985.[28] Recent research has confirmed earlier experiments on hydration aging, but the DOE has yet to seriously address the policy implications, which could be that a repository sited at Yucca Mountain may be incompatible with vitrified glass wastes.[29]

In sum, present data and theory indicate that borosilicate glass may not be the best waste form for preventing the release of radionuclides. There are waste forms that in principle have far lower solubility than glass, but what the NAS study reported in 1983 continues to be true so far as most high-level waste is concerned: these waste forms have not been developed, and there "are at present no substantial development programs within the Department of Energy that are concerned primarily with alternative waste forms."[30] The one exception to this is the plan to develop waste forms other than glass that may be suited to the calcined portion of repro-

25 Bates et al. 1982; Bates 1990.

26 The proposed Yucca Mountain repository is currently above the deep water table at this site. In contrast, many proposed repository sites around the world (including all hard-rock sites previously proposed in the U.S.) would be saturated with ground water. There is some uncertainty about whether the water table at Yucca Mountain might rise in the future and cause water or steam to be present in a repository constructed there.

27 Makhijani 1991.

28 Makhijani and Tucker 1985; Makhijani et al. 1986, 1987; Makhijani 1989; Saleska and Makhijani 1990; Makhijani 1991.

29 Bates 1990.

30 U.S. National Academy of Sciences 1983, p. 82.

cessing waste from naval reactor spent fuel at Idaho National Engineering Laboratory. (Calcining converts wastes from liquid to powdered form for temporary storage.)

REMOVAL OF WASTES FROM TANKS

It may be difficult to empty some storage tanks in order to process the contents into forms suitable for long-term storage. This is a particular problem for the older single-shell tanks at Hanford, where, as mentioned above, much of the waste has been crystallized to a solid salt cake in part due to problems of leakage from tanks. This, however, means that in order to remove the wastes from the tanks they would essentially have to be mined out, a process that would probably have to be conducted remotely because of the high radiation fields. Moreover, the tanks at Hanford contain mixtures of explosive and volatile organic chemicals whose composition is uncertain.

It is because of such difficulties that the DOE has seriously considered the option of simply leaving some of the waste in the tanks indefinitely. As discussed in a 1987 Environmental Impact Statement by the DOE,

> The objective of the in-place stabilization and disposal alternative is to immobilize and stabilize . . . wastes at Hanford and dispose of the waste or provide enhanced protection by isolation from ecosystems using a protective barrier and marker system.
>
> . . . Wastes in single-shell tanks would be dried and the tanks filled with a suitable material to limit future subsidence and provided with interim systems for heat removal as needed. . . .
>
> Although in-place stabilization and disposal would be a permanent disposal action, and retrieval would not be contemplated, the fact that waste has been so disposed of does not preclude future generations intentionally removing the waste (although with some difficulty) for resource recovery or to effect enhanced disposal by some other means if either ever appears warranted.[31]

In the end, the DOE's preferred alternative deferred a decision on what to do with the single-shell tank waste, while choosing the vitrification option for double-shell tank wastes. In making this decision, DOE acknowledged that because single-shell tanks contained more radioactivity, it appeared "counter-intuitive to treat and dispose of double-shell tank wastes before treating and disposing of single-shell tank wastes." DOE stated, however, that since the wastes in single-shell tanks were largely

31 U.S. DOE 1987, p. 3.23.

solidified, "the potential risk from leakage from the double-shell tanks might be greater than from the single-shell tanks."[32]

As a result of an ongoing and extensive re-analysis of the situation of the Hanford tanks, which was stimulated by concerns about tank explosion risks, plans for the disposition of wastes at all Hanford tanks are again in some degree of flux. Recent estimates were that vitrifying single-shell tank waste in addition to double-shell wastes would increase disposal costs by about 170 percent.[33]

The difficulty with removing waste from tanks is less likely to be an issue at newer commercial reprocessing facilities like those in France and the U.K., where the wastes are essentially kept in their initial acidic liquid form and everything remains in solution. The addition of chemicals to neutralize wastes and precipitate some of the materials is the source of the specific problems at Hanford discussed above.

---

32 U.S. DOE 1987, p. 3.35.

33 Official cost estimates are about $15 billion for double shell tanks, and $40 billion if single-shell tanks wastes are included. (Grygiel 1991).

Chapter 7
Warhead Dismantlement and Plutonium Disposal

THE BREAKUP OF the Soviet Union sent a wave of fear around the world over the possibility that control over the 30,000 Soviet nuclear weapons might be lost and that the weapons might be sold by the breakaway states or dissidents for desperately needed cash. A flurry of activity began that was directed towards bringing the warheads to centralized storage locations and disabling them. The U.S. Congress gave a boost with a $400 million appropriation to assist the Soviet program. The process of warhead dismantlement and destruction in the U.S. and Russia took on new meaning amidst hope that it would grow and go far in ridding the world of nuclear arms before its momentum was spent. Simultaneously, there was concern that the retirement and dismantlement process would go too fast, with warheads and materials piling up at unprepared sites, inviting environmental catastrophe by accidental criticality or fire.[1]

Nuclear Warhead Dismantlement[2]

Neither the Intermediate-range Nuclear Forces (INF) Treaty nor the Strategic Arms Reduction Treaty (START) requires the destruction of nuclear warheads or the removal of fissile nuclear materials (highly enriched uranium and plutonium) from the weapons supply stream. The focus there was on elimination of delivery systems.

However, it is finally being recognized that the secure control and elimination of nuclear warheads and nuclear materials is central to nuclear disarmament and to halting nuclear proliferation. Steps may continue in this

1 See Makhijani and Hoenig 1991; Keeny and Panofsky 1992, p. 3; Schneider 1992.
2 See Taylor 1990; Federation of American Scientists 1991a and 1991b; Keeny and Panofsky 1992, p. 5.

direction without any formal international agreement between the nuclear weapon states, but on a reciprocal, unilateral basis. The first such moves were made in September 1991 by President Bush and President Gorbachev in reciprocal pledges, each taken unilaterally without negotiations, to remove and destroy thousands of tactical nuclear warheads. Further, at a summit meeting in June 1992, Presidents Yeltsin and Bush agreed in principle to deep cuts in their respective strategic nuclear arsenals by the year 2003.

Warhead elimination involves a sequence of related steps, including disabling, tagging, transportation, storage, dismantlement, and disposing of the highly enriched uranium and plutonium. Existing Soviet dismantling capacity is 1,500 to 4,500 warheads per year, whereas U.S. capacity is 2,000 to 4,000 annually. However, the capacity of Russia to dismantle nuclear warheads under the prevailing conditions of economic and political upheaval may be far less than the theoretical potential, as with other Russian industries.[3]

It is important that verification go along with and be an integral part of the elimination process. (Unfortunately, there is presently no joint verification of the tactical warhead elimination activities resulting from the Bush-Gorbachev initiative because the Bush administration will not agree to reciprocity. The purpose of verification is to give reasonable assurance that there is no cheating and that warheads and fissile materials are what and where they are claimed to be. Regular warhead verification inspections should be carried out by bilateral or multilateral teams from countries involved in the agreed-on reductions, as well as third parties, such as the International Atomic Energy Agency (IAEA) or non-nuclear weapons states. There could also be international control or U.N. control of warheads and nuclear weapons materials.

The following scenario illustrates how a thoroughly verifiable dismantlement program might work.

As a first step, the host country would declare numbers and types of warheads to be destroyed. Disabling the warheads, so that they cannot be detonated, can be done simply by removing tritium reservoirs, special batteries, electronic firing units, or environmental-sensor arming devices. This safety and security precaution probably would be taken by the host country alone to protect design secrets.

---

3 Thomas Cochran, NRDC, Washington, D.C., personal communication with A. Makhijani, spring 1992; John Large, Large and Associates, London, personal communication with K. Yih, spring 1992.

7.1. All the warheads in the U.S. nuclear arsenal. This field of ceramic nose-cones represents, in miniature, all the nuclear warheads in the U.S. arsenal. Sculpture installation "Amber Waves of Grain," by Barbara Donacky, Boston Science Museum, Boston, Massachusetts, USA, 1985. Photo by Robert Del Tredici, from *At Work in the Fields of the Bomb* (Harper & Row, New York, 1987).

The warheads would then be "tagged" and sealed so that they could be authenticated after transfer to a central site. The tagging could be done by joint teams or by the host country alone. A tag provides a permanent "fingerprint" that can be verified later. Tags can be simple, such as a dab of paint with suspended glitter particles or a surface micrograph taken at a marked spot on the warhead or warhead canister to record irregularities. Each type of warhead has a set of distinctive physical characteristics, such as gamma ray emissions and neutron emissions, which can be ascertained without dismantlement. It would thus be possible to check that warheads were of the type stated by comparing their gamma-spectrum fingerprints, for example, with an agreed standard. Verification would be done by sampling. Finally, to apply a tamper-proof seal, fiber-optic cables can be wrapped and secured around the warhead canister and crimped to provide a unique optical pattern when one end is illuminated. This technique is commonly used in IAEA safeguards.[4]

Transportation of warheads to a central storage site would be done by means of safe secure tractor-trailer or train, like those routinely used by

4   Federation of American Scientists 1991a.

the U.S. Department of Energy.[5] Physical protection of the interior of the storage site would be carried out by the host country, with a multilateral force or UN peacekeeping force around the perimeter.

At various stages it would be desirable to check that warheads or warheads in canisters are genuine and not dummies. A check of warheads entering a storage facility or dismantlement facility can be done by passive or active neutron assay or by gamma ray scanning to be sure that fissile materials are present. If the warhead is in a canister, low-resolution X-rays can be used to determine a warhead configuration without revealing design information.

Warheads would next be shipped to the dismantlement facility. The warheads would be weighed and measured, and once the receipt were jointly verified, they would be taken apart by the host country and prepared for disposal or destroyed. The chemical high explosive would be separated and burned, with unacceptable pollutants removed before venting to the atmosphere. Remaining tritium would be returned or stored under safeguards until it decayed. Small components using radioactive materials, such as neutron initiators, would be separated and stored as high-level radioactive waste. Materials such as deuterium, beryllium, and natural uranium would be returned to the owner country. Remaining electronic components and other materials would be compacted or incinerated before being discharged as waste. All materials declared non-nuclear would be scanned for neutrons or gamma rays to be sure they contained no fissile materials.

The fissile components made of highly enriched uranium and plutonium would be crushed or melted by the host country into ingots so as not to reveal design secrets. The fissile materials would then be passed to the joint team at a specified point in the facility, where they would be assayed and weighed to be sure they correlated with the numbers and types of warheads shipped into the facility.

From the dismantlement facility the fissile materials would be put in safe and secure interim storage, either on site or at another facility, where they would stay subject to third-party verification until long-term disposal were decided by the involved parties. At this point, all nuclear weapons activities in the host country would have halted, and there would be a move toward "full-scope" IAEA safeguards on all nuclear activities, with the right to "special" challenge inspections of suspicious sites. However, the condition of full-scope safeguards should not be allowed to slow warhead destruction.

5   As discussed, for example in Cochran et al. 1987a, p. 41.

A prototype dismantlement facility having a capacity of 2,500 warheads per year would have average daily outputs of 160 kilograms of highly enriched uranium (93 percent uranium-235) and 32 kilograms of plutonium. In addition, the facility would have storage capacity for 800 additional warheads in canisters. It is assumed on average that a U.S. warhead contains 20 kilograms of highly enriched uranium, 4 kilograms of plutonium, and 4 grams of tritium and has an average weight, excluding guidance package, re-entry vehicle, and shipping container, of about 350 kilograms, half of which is chemical high explosive. The total weight of material shipped into the facility per day would be less than 10 tons.

Disposal of Plutonium from Warheads[6]

The quantity of separated plutonium worldwide is about 120 tons in presently-civilian programs and about 240 tons in weapons inventories. Another 200 tons of civilian plutonium are scheduled to be separated in the 1990s as part of presently-civilian programs.

The options for disposal of this plutonium must be judged in light of the magnitude of the problem: the separated plutonium is enough for 100,000 nuclear weapons and a far larger number of plutonium dispersal devices made with conventional explosives (plutonium dispersal devices, or radiation bombs, are discussed in the next chapter).

With the end of the Cold War, the options for plutonium that are being widely discussed cover a range of temporary and permanent measures with various levels of technological sophistication:

1. Monitored and secured storage of plutonium for an indefinite period
2. Fabrication of plutonium into mixed-oxide (MOx) fuel to be used in commercial power reactors
3. Fissioning of plutonium in an accelerator or a nuclear reactor
4. Deep geologic disposal or sub-seabed disposal of plutonium (this might entail mixing it with existing high-level wastes and vitrifying the mixture into glass or ceramic form for disposal)
5. Launch of plutonium into the sun
6. Destruction of nuclear warheads in an underground nuclear explosion

The objective of policy should be to dispose of the material as quickly as possible, while keeping adverse health, environmental, and political conse-

---

6   See Feiveson 1992; Bloomster et al. 1990; Albright and Feiveson 1988; Feiveson 1989, p. 69; Cochran et al. 1987a, p. 41; U.S. Congress OTA 1989; Hebel et al. 1978, p. S114.

quences to a minimum. The overriding goal should be to make the plutonium unattractive and inaccessible for weapons purposes by some combination of technical and political means. The various options for storage or
disposal differ with respect to the recoverability of plutonium. The first
option, for example, is evidently not a "disposal" method as such. It keeps
open the possibility that plutonium will be re-used to make nuclear
weapons.

MONITORED SURFACE STORAGE

Monitored storage of plutonium from dismantled weapons may be
required for a long period if other modes of disposal cannot be agreed on
or if they take a long time to implement. International plutonium storage
was an element of the early Acheson-Lilienthal Plan, and progress towards
nuclear disarmament could see a revival of this idea. Storage could be centralized, or a number of different storage facilities in various locations
could be constructed.

A storage facility probably could be located with the least difficulty at
an existing secure weapons site, such as the Pantex assembly plant in the
U.S or the Chelyabinsk-65 site in Russia. If so, the facility would have to
be physically separate from other facilities at the complex. Further, the
potential environmental problems of storing large quantities of plutonium
at a single site must be carefully considered, especially in densely populated or agricultural regions.

Such a facility would be a high-security vault for plutonium in oxide or
metallic form. It would be designed with built-in safety measures against
accidental criticality and fire. The advantage of storing plutonium in
metal form is that it would require no further processing, since this is the
form in which it is used in weapons. The risk, however, is that it would
be ready for reassembly into nuclear warheads. The oxide form poses its
own problems. Plutonium metal must first be converted to oxide. This
can be done by first converting it to plutonyl nitrate by dissolution in
nitric acid and then calcining it into a powder. Alternatively, it can be
converted directly into oxide in a controlled oxygen furnace. Existing conversion facilities in the U.S., such as the plutonium finishing plant at
Hanford, (and probably also in Russia) are old and environmentally suspect. Also, plutonium oxide, being a powder, can be more easily used to
make "radiation bombs" or dispersal weapons (see next chapter).

Such a facility also would have to be open to verification through bilateral or multilateral inspection, as well as to continuous monitoring by a

third party, such as a special U.N. agency. Resident inspectors would have free access to the facility to carry out intensive verification measurements using weight, high resolution gamma ray spectroscopy, and calorimetry. The plutonium would be stored in standardized containers. Viewing windows allow inspectors to count items to assure that the vault has not been accessed without notice, and a video surveillance camera provides a continuous record of activities in the vault. Because much of the verification activity would be carried out remotely and involve automated handling and on-line monitoring, special measures would need to be taken to assure that data were not being falsified.

A monitored storage facility for 50 tons of plutonium has an estimated capital cost of $170 million (in 1990 dollars) with an operating cost of $28 million per year.[7]

As part of a treaty or agreement, measures would have to be included to respond to attempts, successful or not, to reverse the storage process and remove plutonium from the facility for unauthorized purposes. The response might be to impose international sanctions against the violating country.

The possibility of sabotage or theft might lead many to prefer a more permanent and irreversible form of storage or disposal. The other options reviewed fall into this category.

FABRICATION INTO MOx FUEL FOR COMMERCIAL REACTORS

Perhaps the most controversial disposal option is using the plutonium from weapons as fuel in nuclear power reactors.

Already, some plutonium recovered from spent commercial reactor fuel has been used as fuel by France, Germany, and Japan in their light-water power reactors. Now that uranium is plentiful, large-scale breeder programs have been postponed beyond the first quarter of the next century. Moreover, breeder reactors in France have proven expensive to build and difficult to operate. France's Superphenix has been shut down due to technical problems. Thus, for the interim future, any use of plutonium as fuel for power reactors will have to be in conventional reactors.

Plutonium that has been contracted for separation through the year 2000 at reprocessing plants in France, Great Britain, and Japan will increasingly be used in existing light-water reactors by the owner coun-

7 Bloomster et al. 1990, pp. 12, 13.

tries in Europe and by Japan, which has a particularly ambitious plutoni-um-fuel program. Russia, on the other hand, still wants to set aside large stocks of plutonium for developing breeders.

Use of plutonium fuel in light-water reactors is not economical at pre-sent compared to the "once-through" use of low-enriched uranium fuel, even if the plutonium is "free" (that is, not counting the cost of reprocess-ing or of warhead dismantlement). The reason for this is that the cost of fabricating mixed-oxide (MOx) fuel from plutonium (4 to 5 percent) and natural uranium is very high due to the greater radiation hazards associat-ed with handling plutonium; in comparison, the price of uranium is low and is likely to remain so.[8] Most recent German data place the cost of MOx fabrication at over $2,000 per kilogram of heavy metal, some six times the fabrication cost of low-enriched uranium fuel.[9] Assuming MOx fabrication costing $1,300 to $2,000 per kilogram, the cost of uranium would have to rise to $123–$245 per kilogram to equal it, even if the plu-tonium were free.[10]

A decision to burn weapons plutonium instead of uranium in commer-cial power plants would add to the surplus of separated plutonium build-ing up in Europe and Japan from commercial reprocessing, because, for safety reasons, at most one-third of the core of a current-design light-water reactor can be loaded with MOx fuel at any time. An annual MOx reload to a 1,000-megawatt-electrical light-water reactor could contain at most 0.4–0.5 tons of plutonium, depending on the percentage of fissile isotopes. In addition, MOx fabrication capacity worldwide is extremely limited.

MOx fuel is not licensed for U.S. power reactors. Any future attempt to burn weapons plutonium in U.S. power reactors would probably focus on a small number of reactors designated for that purpose or on new reactors specially designed to take a full core of MOx fuel. In the former Soviet Union, the continuing public reaction to Chernobyl is likely to limit any experimentation with MOx fuel.

Using plutonium as fuel on a large scale would be difficult to safeguard and would involve a high risk of diversion. In the case of plutonium from weapons, there would be a regular traffic of plutonium oxide from dis-

8   The spot price of uranium currently is about $8 per pound of uranium oxide, and it is likely to go lower if Russian weapons uranium enters the commercial market. (Feiveson 1992, p. K-6.)

9   *Nuclear Fuel,* January 26, 1992.

10 Feiveson 1989, p. 69.

mantlement and storage sites to fabrication facilities and reactors, with the risk of attack along transportation routes. In addition, plutonium separated in Europe from power reactor spent fuel would be shipped out in oxide form from commercial reprocessing plants and transported not only within Europe, but by ship to Japan.

These activities would involve hundreds or more shipments of many weapons-worth of plutonium annually. The same level of protection is required for all plutonium, regardless of grade, since all grades are usable in nuclear explosives and in plutonium dispersal devices. Armed transports, such as those used by the U.S. Department of Energy for shipments of weapons materials and warheads, provide layers of protection and maintain continuous radio contact between shipper and receiver. A 1984 shipment of 250 kilograms of plutonium from France to Japan required continuous U.S. military tracking and surveillance at a cost of millions of dollars.[11] According to current plans, future shipments to Japan from Europe are to be accompanied only by a lightly armed escort ship.

Physical protection measures would be taken under national and international auspices, with increased oversight of the U.N. Security Council and the International Atomic Energy Agency. The IAEA has issued minimal guidelines on the physical protection of nuclear materials, and countries are bound by the Convention on the Physical Protection of Nuclear Materials. Requirements for armed guards, quick response to attempted theft, and contingency planning need further strengthening.

Thwarting the diversion of significant quantities of plutonium (8 kilograms) is the target of IAEA safeguards. However, accounting for materials at facilities that handle large quantities of bulky materials such as plutonium oxide is a difficult task involving uncertainties of 100 kilograms or more. Although containment and surveillance can be relied on, they are less quantitative than materials accounting.

The problems with safeguarding plutonium underscore the dangers inherent in promoting it for either commercial or weapons use. Once plutonium oxide is diverted it could be converted into a weapon in a short time (days or weeks).

IRRADIATING PLUTONIUM IN AN ACCELERATOR

One possible scheme, in preparation for final disposal, would be to irradiate plutonium in a particle accelerator beam. Ideally, this process could trans-

11 Spector 1985, pp. 237, 238.

mute all the plutonium into non-fissionable products. This would have a result similar to irradiation of the plutonium in a nuclear reactor. Accelerators chosen for irradiating plutonium could be ones producing intense beams of either neutrons or gamma rays. To be effective, the accelerator must be able to condition at least a few tons of plutonium yearly for further disposal.

The accelerator probably could be built using currently feasible accelerator technology. A likely candidate is a linear accelerator ("Linac") that sends a beam of protons, each with an energy of 2 billion electron volts (GeV), into a thick lead target. The collision of one proton with the lead produces a shower of some 50 neutrons of various energies, which then impinge on encapsulated plutonium targets to produce fissions. The accelerator would be a mile long and have a beam current of some 300 milliamps. This Linac would produce about 3×10^{27} neutrons per year, enough to completely fission 1.2 tons of plutonium.

Plutonium also could be irradiated and transmuted in the neutron flux of a fast neutron reactor. A U.S. research and testing reactor that can perform this task on a small scale is the 400-megawatt Fast Flux Test Facility at the Hanford site, which was built for the test irradiation of fast breeder reactor fuels, materials, components, and systems.

The disadvantages of these approaches that rely on neutrons, and to some extent all transmutation techniques, are:

- Neutrons create induced radioactivity in equipment, and large quantities of radioactive waste would be produced.
- The potential for diversion and attendant security problems is increased by the fabrication of targets or fuel rods and other processes involving plutonium.
- Not all the plutonium can be transmuted, so significant residues of it would remain to be disposed of.
- Long-lived fission products requiring disposal would be created.

DEEP GEOLOGIC DISPOSAL/SEABED DISPOSAL

Deep geologic disposal of spent nuclear reactor fuel is planned for most countries. Likewise, the U.S has already built the WIPP underground repository in New Mexico specifically for disposal of plutonium-contaminated military transuranic wastes, although whether it will actually be allowed to open for that purpose remains in doubt.

If plutonium is mixed with long-lived fission products from reprocessing plant waste and encapsulated, it could be disposed of like spent fuel. Thus, several barriers would be put around the plutonium to thwart diversion into weapons use: high-level radioactivity that would require additional shielding for handling; long-term, nearly irretrievable storage in a geologic repository; and the need to separate plutonium from fission products and the encapsulating material.

Plutonium could be diluted with military and civilian liquid high-level wastes, where they are available in storage, and the mixture vitrified and stored for an indefinite period. This would avoid long-term storage in metal or oxide form. The rate of disposal would be determined by the capacity of the vitrification facility and limitations on plutonium concentration. Ideally, the concentration of plutonium in the glass (or other diluting medium) should be so low that it would be more expensive and difficult to re-extract it than to recover plutonium from spent fuel. Permanence of disposal would be effectively ensured by low plutonium concentrations (because several canisters would have to be diverted in order to obtain a quantity of plutonium sufficient for a weapon) and by the glassification process itself (because glassified waste is difficult, expensive, and hazardous to reverse).

Canisters containing the glassified plutonium and dilutants would be assayed, sealed, and stored at the preparation site since there is currently no geologic repository for high-level wastes. Extended storage on-site is preferable to transportation until such time as a long-term solution is decided upon for all high-level waste.[12]

The technology for borosilicate glass is developed, but areas such as the incorporation of plutonium into the glass and repository design to ensure safety need further careful consideration. Also, borosilicate glass is a durable waste form only under specific hydrogeologic conditions, thus care must be taken that any long-term geologic repository is compatible with it. For example, the Yucca Mountain site in Nevada may not be a suitable one, if water enters the site under certain conditions.[13]

An alternative to geologic disposal is sub-seabed isolation. Canisters would be dropped into the ocean floor to depths of a few tens to 100 meters. The placement and recovery of waste canisters appears achievable with existing deep ocean technology. Further research is still required to clarify uncertainties about the breaching of the containers and the migration of radionuclides within ocean sediments.[14]

---

12 For an extended discussion of nuclear waste disposal problems in the U.S,. see Makhijani and Saleska 1992.

13 Makhijani 1991.

14 Hollister et al. 1981; U.S. Congress OTA 1986.

LAUNCHING PLUTONIUM INTO THE SUN

The once-futuristic option of placing packaged plutonium into earth orbit and then launching it on a trajectory into the sun continues to be discussed and is theoretically feasible with current technology. The technical feasibility depends on having a reliable space flight system with adequate payload and a plutonium capsule that assures no breach of containment even for a worst-case abort or launch-pad explosion.

Serious drawbacks to this option are its cost and the threat to health and the environment from plutonium dispersal accidents and even criticality accidents. Although sun disposal may be the most irreversible option, it is also the most expensive, substantially more expensive than geologic disposal, for example. The cost estimate of sun disposal in a 1982 NASA study is on the order of $200,000 per kilogram of plutonium. A launch would involve hundreds of kilograms of plutonium at one time, perhaps ten times the plutonium contained in one multiple-warhead intercontinental ballistic missile (ICBM). The 1982 NASA study points out that the package must not only be designed to resist rupture from launch pad explosions and terminal velocity impacts due to launch failures, but also melting due to reentry heating and rupture due to deep ocean submergence. It would be difficult and very costly to make a container that would not break under such extreme conditions.

The public response to the threat to health and the environment is already evidenced in the U.S. by uproar over the launch of a plutonium-238-powered satellite for Galileo space missions and by the wide adverse reaction to Pentagon plans for the development of nuclear-powered rockets for ballistic missiles and Star Wars.

UNDERGROUND NUCLEAR DETONATION

The idea of disposal of warheads in a large underground nuclear explosion has gained a greater degree of acceptance and support in the former Soviet Union among some officials and weapons scientists than in the U.S. In the Soviet Union, "peaceful" underground nuclear explosions have long been used in oil and gas recovery, and currently a Moscow company, the International Chetek Corporation, is proposing to sell underground nuclear explosions for commercial applications, including the disposal of toxic wastes.

In this option, a number of warhead "primaries", or quantities of plutonium from dismantled warheads and other sources, would be transported from secure storage and placed in an underground hole at a nuclear test site. Then, a single 100-kiloton warhead detonated in the hole would vaporize the plutonium and fix it in the molten glass mass created by the explosion. In one Chetek proposal, 5,000 or more warheads placed in a large underground cavity would be destroyed with one 100-kiloton explosion.

A more modest approach would use technology developed over 45 years of nuclear testing and place the warheads to be destroyed in underground shafts. U.S. weapons testing shafts are 600 to 5,000 feet in depth and from 3 to 12 feet in diameter. Placing five additional warheads in each hole would require over 3,000 detonations to destroy a stockpile of 20,000 warheads; 50 warheads per hole would require more than 300 detonations. The magnitude of this project can be understood by comparison to the total of some 730 U.S. underground weapons tests.

The detonations would have to be carefully monitored to verify that they were for purposes of disposal and not to test new warhead designs. Each nation could dispose of its own weapons, with observers from other countries and the United Nations to verify agreed-on procedures. If a long wait were likely before detonation, the warheads would have to be disabled first in order to prevent their use if stolen.

The large number and high rate of nuclear detonations that would be required would meet with strong political and public opposition. Risks to health and environment would increase from venting and leakage of radioactivity, possible accidents, and other causes. Such massive underground nuclear explosions would disperse huge quantities of nuclear wastes underground, which would be uncontained in any engineered basic system, such as that proposed for civilian waste. Over the course of hundreds of thousands of years, long-lived radionuclides could eventually seep into groundwater and reach the biosphere. Indeed, this may happen even in the near future, depending on the locations of the explosions.

Chapter 8
Post-Cold War Plutonium

URING THE COLD War, both the U.S. and USSR used the rationale
that the risks arising from nuclear forces on alert were justified
because they ensured peace through deterrence — neither side
would dare play the nuclear adventurist and launch a massive first
strike.[1]

Today's reality is symbolized by a request that would have been in the
realm of the surreal two years ago — Russia has asked to join NATO.
Today, the U.S. government is spending money to prevent Soviet scientists from selling their nuclear weapons expertise to the highest bidder.

In spite of the changed climate, the vast arsenals of weapons of mass
destruction built up during the Cold War continue to present a variety of
security and environmental threats. The challenge for policy today is the
containment and elimination of these weapons, particularly nuclear
weapons. There are many dangers that lurk in the nuclear arsenals of the
world. They are:

- nuclear weapons proliferation
- dispersal of plutonium or other radioactive materials by groups or countries that cannot make a nuclear explosive weapon — by means of a
  "radiation bomb"
- radioactivity dispersal due to accidents
- accidental nuclear war

We will consider each of these in turn.

---

1 The term "deterrence" actually had a broader meaning in U.S. strategic doctrine. It
   was related to the idea of "containment" of the Soviet Union. The threat of the use
   of nuclear weapons in any conflict and the threat of all-out nuclear war against the
   Soviet Union were elements of a strategy to prevent the Soviet Union from extend-
   ing its influence to other countries and to develop and maintain, as U.S. National
   Security Memorandum Number 68 put it, "a healthy international community" for
   the U.S.-dominated economic system. For a discussion, see Ege and Makhijani 1982.

Nuclear Proliferation

The end of the Cold War and the U.S. insistence on continuing a policy of "deterrence" means that other countries will continue to find the idea of acquiring nuclear weapons capability attractive. In addition to the direct military appeal, possession of such weaponry appears to command respect and to confer a measure of political power in the international arena. The attention and aid being given to Russia out of apprehension about the destiny of its nuclear weapons and related expertise is an example that surely has not been lost on the people or governments of Third World countries (in spite of the threat of economic reprisals and direct military intervention issuing from quarters such as the United States).

In view of the break-up of the Soviet Union — accompanied by ongoing tensions among republics and ethnic minorities — the possibility of nuclear proliferation within the former Soviet Union is by no means a trivial concern. We have already seen conflicts arise between Russia and Ukraine and between Russia and Kazakhstan over the control of nuclear weapons and other military materiel and infrastructure.

Finally, increasing trade and other frictions between the U.S. and Japan and, to a far lesser extent as yet, between Germany and the U.S. could take turns for the worse. This would give the plutonium stocks from the breeder reactor programs of these countries an entirely new, military significance. As we have discussed, both countries have the materials and technology to make nuclear weapons; they also have the technology to make many kinds of short-, medium- and long-range delivery systems for these weapons, including missiles.

Not only is the demand for plutonium for weapons still present, but the supply of plutonium is substantial, as we have seen. The quantities of plutonium in nuclear weapons that could contribute to the problem of proliferation include the 120 tons or so in the weapons of the former Soviet republics. In addition, as of 1992 there are about 30 tons of presently-civilian plutonium stockpiled at Chelyabinsk, supposedly for the now-stalled Soviet breeder reactor program. There is also a further potential source of plutonium in the spent fuel from nuclear power plants in the former Soviet Union, as well as fuel that will be irradiated in the years to come to generate electricity. At Chelyabinsk-65, for example, separation of plutonium continues at a rate of about 2.5 tons per year.[2] Furthermore, the 600-megawatt breeder reactor at Beloyarsk (near Ekaterinsburg, for-

2   von Hippel 1992.

merly Sverdlovsk) is still running on enriched uranium fuel even after 12 years, apparently for production of plutonium, judging from the fact that one-third of the core is refueled twice a year and the blankets are replaced often. The spent fuel is taken away. Kazakhstan has a similar 350-megawatt reactor in operation.[3] The extracted plutonium could be stolen and/or sold.

The prospect of a black market in nuclear weapons or their parts developing in the former Soviet Union, which has been raised by a number of analysts,[4] seems plausible given the deepening economic crisis, the need of both institutions and individuals to acquire foreign exchange, and the weakening of central control. Indeed, the sale of large quantities of conventional armaments by Russia for purely commercial purposes, as is being proposed by Russian institutions and statesmen, does not portend well for nuclear non-proliferation.

There are also substantial amounts of plutonium in unreprocessed spent fuel in Germany, Japan, and other countries. Of course, as civilian reactors continue to generate power, they also continue to generate additional amounts of unseparated plutonium.

Radiation Bombs

Groups or countries that cannot manufacture nuclear bombs because they lack the know-how or sufficient materials could still wield considerable power and do great damage by spreading plutonium by explosion with conventional explosives. The potential for such threats has long been recognized, although, like the others discussed here, they have been overshadowed by the exigencies of the Cold War.

The first documented discussion of the use of radioactive materials as weapons of war arose during World War II. The U.S. considered the possibility that its enemies would retaliate with radioactive warfare in response to the U.S. use of nuclear weapons; it also considered using such a weapon itself against its enemies.

There was serious discussion in the U.S. of the possibility of radioactive warfare by the Nazis in 1943 and 1944. James Conant, one of the two main scientific advisers who served as a liaison to the political decision-

---

3 Personal communication from John Large of Large and Associates, London, to Katherine Yih, July 15, 1992. Large was in Ekaterinsburg in July 1992.

4 For example, Makhijani and Hoenig 1991 and Broad 1992.

making of the Manhattan Project, wrote a memorandum on the possible effects of the use of radioactive materials in war in July 1943. He concluded that, despite the difficulties of manufacturing sufficient radioactive material, the dangers of handling it, and the logistical problems of dispersing it from the air, "a series of circumstances might enable the Germans to produce in a city such as London a concentration of radioactive solids over areas varying in size from half a square mile to several square miles sufficient to require the evacuation of the population."[5] The document further stated that the main danger from such contamination was not large-scale deaths, since the amount of radioactivity that had to be dispersed from the air would be too large in the context of the war. Rather, the main danger was the contamination of a large part of a city sufficient to require its evacuation.[6]

The U.S. considered the use of radioactive warfare against its enemies as well. Barton J. Bernstein, a historian of the U.S. nuclear weapons program, has pointed out that in May of 1941, seven months before Pearl Harbor, a special American scientific panel proposed that, "as a top priority, the U.S. develop radioactive products for use against the enemy."[7] In 1943 Enrico Fermi secretly proposed using fission products to poison the enemy's food supply. Robert Oppenheimer, who led the scientific effort to build the first nuclear weapon, pursued the plan directly with Edward Teller, and proposed that "we should not attempt [such] a plan unless we can poison food sufficient to kill a half million men, since there is no doubt that the actual number affected will, because of non-uniform distribution, be much smaller than this."[8]

Joseph G. Hamilton, an assistant professor of medicine at Berkeley's Radiation Laboratory, who was experimenting with radioactive products on animals and humans and who worked with Oppenheimer, proposed contaminating large reservoirs of water with a million curies, rendering food supplies "unfit for consumption."[9] In a letter, Hamilton wrote:

> The idea is something like this — if you wished to raid a place and make everybody nauseated, vomiting and incapacitated within a period of 24 hours, how much radioactive material in either [beta]-ray emitter or [gamma]-ray

5 Conant 1943.
6 Conant 1943.
7 Bernstein 1985, p. 44.
8 Bernstein 1985, p. 46.
9 Bernstein 1985.

emitter type is needed? If you wish to get the same effect within a week [or] to keep an area uninhabitable for a month?[10]

The importance of the threat of radiation in military strategy is more clearly seen in the evaluation of the bombings of Hiroshima and Nagasaki made by the U.S. military and in related U.S. planning after World War II to integrate nuclear weapons into U.S. military strategy.

Post-World War II military documents are graphic in describing the way in which the threat of radiation would terrorize people and (hence) lead nations to capitulate. One document described the radioactive lessons to be learned from the atomic destruction of Hiroshima and Nagasaki and particularly from the first post-World War II nuclear tests at Bikini in 1946. The second test in 1946, called Test Baker, was an underwater test that threw up millions of tons of contaminated water and created vast radioactive mists. This "would have not only an immediately lethal effect, but would establish a long term hazard through contamination of structures by deposition of radioactive particles."[11] The report further describes the effect of the nuclear explosion and the subsequent dispersal of radioactivity upon the population of a city. It is particularly eloquent about the effects of the lingering radioactivity:

> We can form no adequate mental picture of the multiple disaster which would befall a modern city, blasted by one or more bombs and enveloped by radioactive mists. Of the survivors in the contaminated areas, some would be doomed by radiation sickness in hours, some in days, some in years. But, these areas, irregular in size and shape, as wind and topography might form them, would have no visible boundaries. No survivor could be certain he was not among the doomed, and so added to every terror of the moment, thousands would be stricken with a fear of death and the uncertainty of the time of its arrival.[12]

The evaluation of the underwater test stated that advance knowledge of the effects of radiation would be particularly effective in complementing the effects of the explosion, providing "psychic stimuli" that were lacking in the air-burst explosions that destroyed Hiroshima and Nagasaki.[13] The evaluation of the bombings of Japan and of the two post-war tests in 1946 at Bikini concluded as follows about the combined power of explosions and radioactivity:

10 Bernstein 1985.
11 U.S. Joint Chiefs of Staff 1947, p. 84.
12 U.S. Joint Chiefs of Staff 1947, p. 84.
13 U.S. Joint Chiefs of Staff 1947, p. 86.

In the face of . . . the bomb's demonstrated power to deliver death to tens of thousands, of primary military concern will be the bomb's potentiality to break the will of nations and of peoples by the stimulation of man's primordial fears, those of the unknown, the invisible, the mysterious. We may deduce from a variety of established facts that the effective exploitation of the bomb's psychological implications will take precedence over the application of the destructive and lethal effects in deciding the issue of war.[14]

It is interesting to note that during the very period in which the U.S. military was making an evaluation that the effects of lingering radioactivity would be among the most deadly as well as the most sapping of morale of an adversary, it was implementing a decision to locate a nuclear test site within the continental United States. However, the U.S. public was fearful of radioactivity from nuclear explosions (quite appropriately, even according to the formerly secret evaluations of the Joint Chiefs of Staff). Locating the test site within continental U.S. limits therefore presented "public relations" problems for the military, which decided to resolve them through what one document called a "reeducation campaign" to reassure the public that it was safe to have nuclear weapons tests "within a matter of hundred or so miles of their homes."[15] Thus, in about the same period that the U.S. military was engaged in reassuring the U.S. public about the safety of tests and, implicitly, of radioactivity, it was considering using the threat of radiation to "break the will of nations and of peoples."

The acuteness of the present danger from plutonium is illustrated by the nature of the technical difficulties that a would-be radioactive bomb-maker would have to overcome, as specified by Conant. They were:

- the difficulty of getting large quantities of radioactive material
- the dangers from the radioactivity of the material to those who would be handling it
- the difficulty of dispersing a large amount of radioactive material from the air.

In the context of surplus plutonium from nuclear weapons, the first difficulty is irrelevant — the material has already been made. The second difficulty is not as great for plutonium as it is for many other radioactive materials (especially gamma emitters), since the alpha particles from plutonium can be blocked by relatively modest shielding. The main dangers of handling and transportation arise from accidental inhalation or, to a lesser extent, ingestion.

14 U.S. Joint Chiefs of Staff 1947, pp. 86–87.
15 This question is discussed in detail in IPPNW and IEER 1991, Chapter 4.

There remains some difficulty in dispersing plutonium into the environment, since it is used in metal form in nuclear weapons, a form not very well suited to dispersal. But since plutonium is pyrophoric, radiation bombs could be designed that would use this property to convert plutonium into fine particles and achieve dispersal.

Accidental Dispersal

Plutonium in nuclear weapons can be dispersed into the environment accidentally without a huge nuclear explosion. One route to such dispersal is the accidental partial detonation of the conventional explosives in the warhead. Nuclear warheads have powerful conventional explosives built around the plutonium trigger. When the nuclear warhead is to be exploded, an electrical signal sets off the conventional explosive. This compresses the plutonium core in the warhead and makes the plutonium into a supercritical mass. If only a part of the conventional explosive is accidentally detonated, or the whole detonated unevenly, the explosion will likely fail to trigger the nuclear blast, but will disperse the plutonium compounds widely.

Stray radio waves or static electricity can, under special circumstances, trigger the electronic ignition devices. The accidental detonation of explosive due to stray radio waves is known as the "hazard of electromagnetic radiation to ordnance" (the H.E.R.O. effect). Nuclear weapons and other devices containing such explosives are shielded against stray radio waves and static electricity, and considerable precautions are taken to prevent accidental detonations. Despite these precautions, there have been accidents involving conventional explosives of the kind used in nuclear weapons. For instance, in West Germany in 1985, the solid rocket fuel motor of a Pershing II missile exploded due to such an effect, killing three U.S. soldiers and severely injuring seven others.[16] Fortunately, it was not armed with a nuclear weapon at the time.

Accidental Nuclear War

Accidental nuclear war has been one of the main dangers from nuclear arsenals, particularly since missiles (which, unlike bombers, cannot be recalled) became a central part of the arsenals of the nuclear weapons powers. A considerable amount of effort and technology has been devoted to detecting

16 May 1989.

incoming missiles and to distinguishing these from natural phenomena or non-nuclear objects. Nonetheless, false alarms have been frequent. Some of them have caused increases in the alert levels of nuclear forces, which constitute the first steps towards the initiation of nuclear war.[17]

With the end of the Cold War, the danger of accidental nuclear war has declined in some ways. For instance, bombers have been taken off alert. More important, the danger that false alarms will trigger a rapid launching of missiles due to these forces being on hair-trigger alert has been much reduced. However, the danger of accidental nuclear war continues to be one of the main problems from the nuclear arsenal.

While the probability that false alarms may trigger an accidental war has declined, it is still possible that an accidental firing of a nuclear missile due to malfunctions or human error could escalate into a large-scale affair, if nuclear weapons are armed and ready to go at short notice. The terrible tragedy of a single accidental or deliberate nuclear explosion could turn into utter global disaster.

17 May 1989, pp. 254–257.

Chapter 9
Summary and Recommendations

A S A KEY ingredient of nuclear weapons and a presumed "magical" source of energy, plutonium has been one of the main currencies of military and political domination since 1945. Since that time, nuclear weapons have not only provided a means of international domination but they have vested an inordinate amount of power in an elite group of bureaucrats in the nuclear-weapon states. Alvin Weinberg, former director of the U.S. Oak Ridge National Laboratory, clearly recognized this development in the following statement:

[N]uclear weapons have stabilized at least the relations between the superpowers. The prospects of an all out third world war seem to recede. In exchange for this atomic sense, we have established a military priesthood which guards against inadvertent use of nuclear weapons, which maintains what *a priori* seems to be a precarious balance between readiness to go to war and vigilance against human errors that would precipitate war. . . . The discovery of the bomb has imposed an additional demand on our social institutions. It had called forth this military priesthood upon which in a way we all depend for our survival.[1]

Regrettably, the decisions, assumptions, and arguments made by the international nuclear priesthood have been flawed and have proven disastrous for many people. In the case of plutonium, the dangers to public health, environmental quality, and world security posed by its production and existence have not ended with the Cold War. Indeed, plutonium presents a number of new, acute problems in this era, which urgently demand resolution.

---

1 Weinberg 1972.

Summary of Findings

1. Plutonium is one of the deadliest substances known.

Plutonium is an alpha-emitting transuranic element. Of the possible routes of entry into the body, the most common and most dangerous is through inhalation. In addition to irradiating lung tissue, plutonium is gradually transported to other organs, in particular, liver and bone. Once an alpha-emitter is inside the body, its radiation can cause genetic mutations and cancer with greater potency than gamma or beta radiation of the same energy. Recent research on transmitted chromosomal instabilities in mouse hematopoietic stem cells and sister chromatid exchanges in hamster ovary cells suggests that alpha radiation may be even more dangerous than previously thought. Experiments with beagle dogs suggest that about 27 millionths of a gram of insoluble plutonium would be sufficient to cause lung cancer in an adult human being with virtual certainty, with significant risks probably associated with far lower doses.

2. Many countries possess large quantities of military plutonium.

The five nuclear weapons powers have huge amounts of plutonium in their nuclear weapons. The former Soviet Union is estimated to have about 120 tons. The United States has about 110 tons in its weapons program. France, Britain, and China have smaller amounts in their weapons programs — six tons, five tons, and 1.25–2.5 tons, respectively. In addition, Israel and India are thought to have lesser quantities of military plutonium (under one ton). Pakistan may also have some plutonium production capacity, but as of the end of 1991 its weapons program was based primarily on highly enriched uranium. Three to five kilograms of plutonium are required to make one nuclear weapon, although a crude weapon requires more.

3. Nuclear weapons can be made from civilian plutonium possessed by a number of countries.

France, the U.K., Russia/CIS, the U.S., Germany, Belgium, Japan, and India all possess reprocessing capability or substantial quantities of plutonium for their civilian power programs. However, this plutonium can also be used to make nuclear weapons. All these countries also possess the technology, such as bombers and missiles, for delivering nuclear weapons over considerable distances. In addition, Pakistan, Argentina, and Brazil have or had programs in various stages of implementation to develop civilian reprocessing capacity.

4. A large quantity of highly radioactive liquid waste is generated during plutonium production (reprocessing), much of which is stored in tanks.

Plutonium production requires the separation of plutonium and uranium from fission products and from each other. This results in highly radioactive wastes, which are typically stored in large tanks. The volumes in the U.S. are far larger than those in other countries because the wastes in the U.S., which are acidic on discharge from the reprocessing plant, were neutralized with large volumes of sodium hydroxide. In Europe, wastes have generally been stored in acidic form. Acidic and alkaline wastes each pose hazards. For example, alkaline wastes are more difficult to characterize and monitor due to the variety of chemicals that have been added and due to their non-homogeneous nature. They are also more difficult to solidify into glass for final storage. Acidic wastes, being highly concentrated, are much more radioactive than wastes that have been made alkaline, thus they generate much more heat and require constant cooling.

5. High-level wastes have also been discharged directly into the environment.

In the former Soviet Union (and to a far lesser extent in the United States) highly radioactive wastes have been directly discharged into the environment. One area on the shore of Lake Karachay near the Chelyabinsk-65 nuclear weapons production facility in the Ural Mountains is so radioactive that one can get a lethal dose of radioactivity (about 600 rem, or 6 sieverts) simply by standing there for one hour. There are also indications that the Russians continue to put high-level wastes into the environment via deep underground injection.

6. The 1957 explosion (the so-called "Kyshtym accident,") in a high-level waste tank at the Soviet Chelyabinsk-65 site caused a great deal of human and environmental damage.

The explosion contaminated 15,000 square kilometers. More than 10,000 people were evacuated. Information about this accident was suppressed by Soviet officials and the U.S. government for more than three decades. The official estimates released in recent years by the central ministries claim that there are no detectable excess cancers. However, this is unlikely given the high levels of contamination reported. At least for the more heavily contaminated areas, considerable increases in cancer likely occurred in relatively small communities. Our estimate of fatal cancers among the over

10,000 people who were evacuated ranges from 100 to 200. Preliminary studies by some local health researchers indicate an excess risk of leukemia in the region. In addition, we suspect that large numbers of workers suffered high radiation doses during clean-up of the site.

7. High-level wastes can catch fire or explode in various ways.

High-level radioactive wastes sometimes contain potentially combustible organic compounds such as ferrocyanides, and virtually all wastes generate hydrogen, a flammable gas. This means that failure of ventilation or cooling systems, a spark, or the presence of hot spots could initiate a fire or explosion. The composition of wastes, and thus the probability and mechanisms of possible explosion, vary from one country to another, from one reprocessing operation to the next, and even from tank to tank.

8. The solidification into glass of high-level wastes poses some difficult problems in the United States.

The technology proposed for long-term management has been to mix the wastes with molten glass and cast them into large glass logs in a process called "vitrification." Vitrification plants in several countries are in operation or under construction. The longest operating plant is at Marcoule in France. The U.S. vitrification program is faced with considerable difficulties, notably at Hanford due to the very complex nature of the waste. There are as yet no safe ways to effectively pretreat these explosive wastes at Hanford so as to put them into a form suitable for vitrification.

9. There is as yet no operational method for disposal of high-level waste.

Plans for the construction of repositories for high-level radioactive waste have encountered problems in virtually all countries due to poor methods for site selection and vigorous public opposition. In the United States, the search was confined to one site by political fiat. It appears that this site may be unsuitable for the kind of glass that the DOE has chosen for the vitrification of military high-level wastes.

10. Dismantling weapons will pose considerable environmental risks.

Plutonium from unwanted weapons will pose substantial environmental problems in terms of storage, processing, and use or disposal. Each option, whether it be use as an energy source in reactors or processing and disposal as a waste, poses considerable problems. Moreover, there are as yet no facilities to handle the large quantities of surplus plutonium from the dismantling of U.S. and Soviet weapons.

11. Post-Cold War plutonium poses many serious environmental and security problems.

The inability of the nuclear weapons powers to dismantle large quantities of weapons and to dispose of the plutonium in the short run has created a situation where even the unwanted plutonium is now a serious risk. There are numerous proliferation possibilities, many arising from the disintegration of the Soviet Union. There are risks of accidental nuclear war and of accidental dispersal of plutonium. There are also risks that subnational groups or countries which cannot make nuclear weapons might acquire plutonium for use in "radiation bombs". Such bombs have been considered to be dangerous ever since World War II.

Recommendations

1. The secrecy that surrounds plutonium production must be ended.

The Cold War is over, yet in most countries there is still a great deal of secrecy surrounding plutonium production and even environmental issues such as composition of wastes (with the notable exception of the U.S., for many categories of information). It is unacceptable that such information should be kept secret from the very citizens the nuclear weapons were supposed to protect. Not only is the information necessary to know what radiation hazards people have been exposed to in the past and may be exposed to in the future, but it is needed for verifiable dismantlement and storage of weapons and their components. Verifiable dismantlement is a necessity for security, safety, and environmental protection.

2. Analysis of health data on the people exposed to radiation from plutonium production at the Chelyabinsk-65 complex should be carried out by agencies and individuals independent of the nuclear establishment of any country.

The U.S. Departments of Energy and of Defense have shown great interest in the Chelyabinsk data and have made moves to get access to these data. It is important that an independent assessment of the data, beginning with their validity, be carried out. If the U.S. government is to be involved, whether bilaterally or multilaterally, the Department of Health and Human Services is the U.S. agency under whose auspices the collaboration should be arranged.

3. Further separation of plutonium for military or civilian purposes should be stopped.

Plutonium and the wastes generated from it pose unacceptable security and environmental risks, and further production of this material should be stopped. This includes plutonium in civilian programs, since (unlike low-enriched or natural uranium used in civilian power reactors) it can be used to make nuclear weapons. Given the large quantities of uranium that are available and the high cost and security risks of plutonium separation and use, civilian plutonium production cannot be justified against the risks it poses.

4. Liquid wastes in high-level waste tanks should be solidified and stored on site.

Since high-level wastes in liquid form are explosive under certain accidental conditions, these tanks should be emptied and the wastes solidified into an appropriate waste form. (This solidification poses considerable problems at Hanford, however, where the many types of explosive and combustible materials in the tanks make it difficult and dangerous to empty them.)

5. All warheads should be separated from their delivery vehicles and stored in criticality-safe containers.

All nuclear warheads still designated as part of the arsenals of the nuclear weapons powers should be removed from missiles, ships, and airplanes, stored in containers designed to prevent an accidental nuclear criticality, and shielded from stray electrical signals. This would practically eliminate the risks of accidental nuclear war and accidental contamination. Removing weapons from world-wide patrol would also reduce the threat that near-nuclear-weapons states feel from the nuclear weapons powers, which they use as justification for pursuing their own nuclear weapons programs.

6. Plutonium in existing weapons and civilian programs should be put under secure international control; any transportation of it should be for the sole purpose of final storage under international control or preparation for such storage.

Given the health, environmental, and proliferation risks posed by plutonium, we cannot support plutonium-fueled nuclear power.

The great risks of accidental nuclear war, proliferation, and environmental radioactive contamination would be mitigated considerably by securing control of all plutonium, including that which is not at present in weapons. This material should be stored under international supervision under arrangements similar to those for storing weapons. Thus, the sequestering of warheads would be accompanied by the sequestering of nuclear weapons materials.

7. Plutonium should be treated as a hazardous waste material rather than as a resource.

All plutonium should be treated as a waste material never again to be used. Disposal techniques compatible with this conception should be employed. Such methods should be effective and permanent. At present, mixing plutonium with high-level wastes and vitrifying it at existing vitrification plants might be the swiftest way of accomplishing this. We also recommend careful consideration of other alternatives for waste disposal. An environmental impact statement, both for the U.S. and for global plutonium stocks, is urgently needed for all known options.

Appendix

Some Basics of Nuclear Physics and Radiobiology

Elementary Particles

ALL MATTER THAT we shall discuss (except cosmic rays) can be thought of as made up of three kinds of particles that act as if they are indivisible. The proton is a particle whose weight is known (600 billion trillion would make 1 gram), and its weight is used as the unit in the scale of atomic weights. The proton has a fixed amount of positive electric charge, which is called 1 unit as it cannot be divided. The neutron is slightly heavier and can be thought of as being composed of a proton and an electron. Neutrons are therefore electrically neutral. They are stable in the nucleus but in the free state decay into protons and electrons. The electron is a much lighter particle (approximately 1/2000 of the proton weight), and with a single negative electric charge.

An atom has a central nucleus made of neutrons and protons tightly bound together and an encircling cloud of electrons. The cloud contains the same number of electrons as there are protons in the nucleus, so that the whole atom is electrically neutral (equal numbers of positive and negative charges). The electron cloud is shared with other atoms in chemical compounds, and one or more electrons can be split off, or added, to make positively charged or negatively charged bodies called "ions".

The lightest atomic nucleus, hydrogen, consists of just one proton; one electron orbits around it. The heaviest natural atom, uranium, has 92 protons and 146 neutrons forming its nucleus, and 92 orbiting electrons. The "atomic weight" of hydrogen is thus 1, and that of uranium is 238 (92+146).

The atoms of any one element are not necessarily all exactly alike. All have the same number of protons and the same number of surrounding electrons (hence the same chemical properties), but they may have different numbers of neutrons, and therefore different weights. Thus, there may be several different kinds of atom for any one element, and the different kinds are called the isotopes of that element. The mix of isotopes in a natural element normally stays the same throughout its chemical reactions, but the different isotopic weights make slight differences in the physical properties, which allows partial separation or differential concentration by complex physico-chemical processes. Separation of isotopes, in amounts of grams and kilograms, was done first for the development of the atomic bomb during World War II. Prior to that only trace amounts had been separated in academic research laboratories.

Atomic weights are near to whole numbers. The exact weight of an atom on the atomic weight scale differs fractionally from the number of protons plus the number of neutrons, first because neutrons are very slightly heavier than protons, and second because the inherent "binding energy" makes a slight difference to the final weight.

There are three isotopes of hydrogen. All the nuclei have one proton. H-1 (two of these make a molecule of ordinary hydrogen gas) has only a proton; its atomic weight is 1. H-2 (called deuterium) has one neutron combined with a proton, and atomic weight 2. H-3 (tritium) has two neutrons and a proton, and atomic weight 3. Other elements do not have separate names for different isotopes; each isotope is designated by the symbol for the element and the atomic weight, thus, U-238 is the symbol for the commonest isotope of uranium, and U-235 for the isotope with three fewer neutrons, which forms about 0.7 percent of natural uranium.

Depending upon the ratio of neutrons to protons, the nuclei of some isotopes are stable, but others have varying degrees of instability which causes them spontaneously to split off a particle, or occasionally to break in two ("fission"). A large amount of surplus energy (relative to the tiny size of a nucleus) is available, which appears as the high speed of the emitted particles, or as electromagnetic radiation. The whole process is called "radioactive decay". Each nucleus splits at random, independent of the other atoms, which results in the total number of remaining (unsplit) atoms falling to half in a certain length of time that is independent of the original amount of the isotope present. This time, characteristic of the particular isotope, is called the "half-life". Technically it is easy to measure half-lives. They range from billionths of a second to billions of years.

As a consequence of this process of radioactive decay, the amount of an isotope remaining (for example at a certain time after an episode of contamination) can be calculated from the known half-life. After one half-life, the amount is half; after two half-lives, a quarter; after ten half-lives, approximately one-thousandth; and after 20, approximately one-millionth. Thus, an isotope with a half-life less than two weeks is drastically diminished at one year, but one with a half-life of more than 100 years lasts longer than human civilizations.

Radiation

When radioactive decay was first discovered, three forms of radiation were detected, called alpha-, beta-, and gamma-rays. It is easiest to describe them in reverse order.

Gamma-rays are a form of "electromagnetic radiation," a term that covers an enormous range of rays with different properties. In order of increasing energy, the spectrum includes radio-waves, infra-red rays, visible light, ultraviolet light, X-rays and gamma-rays. All electromagnetic radiation travels at the same speed, the speed of light. Its energy acts as if it is in little packets, each called a quantum or photon. In the case of X- or gamma-rays, each photon has sufficient energy to ionize an atom that it strikes, by knocking an electron out of orbit. Photons of the other types of radiation do not have sufficient energy to do that. There is no sharp distinction between X-rays and gamma-rays. The terms are interchangeable, but "gamma-rays" is commonly used for radiation that comes from the nucleus of radioactive atoms, while the term "X-rays" is used for rays generated in an electrical apparatus made for the purpose. Gamma-rays are emitted in conjunction with emission of alpha- or beta-radiation in radioactive decay and in other types of nuclear reaction.

The commonest type of radioactive decay of nuclei is by emission of fast electrons, usually travelling at a speed approaching the speed of light. This is known as beta radiation. These electrons do not come from the cloud of orbiting electrons, but from the nucleus. The emission of an electron from the nucleus is accompanied by the conversion of a neutron into a proton, so electrical charge is conserved. The atomic weight is not changed appreciably, but the atomic number is increased by one, making the atom into a different element. In most cases there is surplus energy that is emitted in the form of gamma-rays.

Alpha radiation is given off by radioactive decay of some isotopes, mostly the heaviest ones, including uranium, radium, and the artificial element plutonium. The alpha radiation is a fast particle consisting of two protons and two neutrons, which happens to be the same as the nucleus of the helium atom. The atomic weight of the isotope is reduced by four units, and the atomic number by two. Again, gamma radiation is usually emitted at the same time, and the atom turns into a different element.

A less common form of beta-activity is the emission of a positron, which is a particle exactly like an electron, but with a positive unit of electric charge instead of a negative one. As soon as it slows down, the positron combines with any electron it comes close to, and the two are annihilated in a burst of gamma-rays.

The ranges of these three types of radiation are quite different. Alpha radiation, at the energy levels given off by radioactive isotopes, is rapidly stopped by matter. In water or living tissue, the range is only a fraction of a millimeter. Beta radiation is more penetrating (giving up its energy more slowly), and its range is typically a few millimeters in water or tissue, a few meters in air. Gamma radiation is much more penetrating, and depending on the energy, travels through many centimeters of water or tissue. Its range depends on frequency; it is gradually attenuated, like the gradual dimming of visible light travelling through tinted glass or dirty water.

Radiobiology

The common factor in the interaction of alpha, beta, and gamma radiation (as well as X-rays) with living cells is that the rays cause ionization of atoms they hit in their path, as well as knocking on other electrons or atomic nuclei, which also cause ionization as they travel. Ionization of an atom immediately breaks up the molecule (chemical compound) of which it was a part. If a single ionization occurs within a gene, that gene is immediately broken or damaged. However, the total gene material forms only a small fraction of the total volume of a living cell, so relatively few ionization events occur in genes. Ionization in water (which is present in all living cells) forms very reactive ions and unstable compounds of hydrogen and oxygen, which can interact chemically with vital chemical compounds of a living cell and thus may destroy the cell or impair its ability to grow and divide.

Cells can survive and repair many injuries. Failing that, the commonest result of damage is cell death, either at the time of injury or when it next divides. In moderate numbers, deaths of cells are harmless to the whole body, which has great ability to repair damage and remove dead material.

A relatively very rare event is a change in the cell's power of division, letting it escape from the body's normal control and continue to grow and divide. That is the kind of change which can become a cancerous growth, if the normal defenses of the body fail to stop it.

A large single dose of radiation to the whole body, in the region of 1 sievert, causes sickness immediately or within hours or days. The predominant effects at this dose level are acute damage to the intestinal tract, the liver, and the bone marrow. Death may follow within a few weeks. After a whole-body dose of 4 sieverts within an hour or so, it is estimated that approximately 50 percent of exposed healthy adults recover from immediate ill-effects, though their risk of cancer and other long-term problems is increased. The other 50 percent die within days or weeks.

If the radiation is delivered slowly over many days, or if several smaller doses are received at intervals, the human body can survive a total dose several times as high. If the individual survives, many risks to future health are increased, including the risk of cancer. The lethal dose for other species of mammals and advanced plants is comparable to that for humans, mostly somewhat higher but a few lower. Simpler animals and plants tend to tolerate higher doses; bacteria, viruses, and dry spores and seeds, much higher.

Much smaller doses than this increase an individual's risk of getting cancer after a number of years, and there may be no dose so small that the risk is not slightly increased. In connection with the late effects of small doses of radiation, it has to be remembered that we all live in a background of natural radiation, and all life on earth has developed with a varying amount of natural background radiation all the time. The average level (disregarding radon and all man-made radiation, whether from "fallout" or medical X-rays) is about one millisievert a year at sea level. The level increases with altitude. Radon doses due to leakage into houses vary considerably, depending on the composition of the underlying soil and house construction details. Disregarding man-made radiation sources, the general rate of cancer accounts for about 20 percent of all deaths. Some fraction of that risk may be due to the background radiation. Most cancers have other causes, of which some, like smoking, are known but others are unknown.

Leukemia, a cancer affecting the bone marrow and therefore the blood, is the cancer most commonly increased in a population exposed to excessive radiation short of a dose that is itself lethal. Leukemia develops more quickly than the "solid cancers," in 2 to 5 years. Solid cancers develop slowly, typically in 10 years or more.

The biological effects of a given dose of radiation depend on the type of radiation. For X-rays and gamma-rays the different photon energies are not very different in their effect, but more densely ionizing radiations, notably alpha-particles and neutrons, have greater biological effect per unit of absorbed dose. The ratio of a standard dose of X-rays to the dose of another type of radiation that causes the same level of biological damage is called the "Relative Biological Effectiveness" (RBE) of the particular radiation. (The concept was worked out in studying treatment of cancer by radiation, which is why the terminology suggests that biological damage is an advantage.) The RBE for any particular radiation may be different according to which biological effect is being studied.

The RBE of alpha-particles is different for various biological effects, and may range up to 60 for some effects at very low doses. An RBE of 20 is generally used in dose calculations for purposes of compliance with radiation protection standards. Recent work by Kadhim et al. (1992) at the British Medical Research Council's Radiobiology Unit indicates a different kind of effect for certain genetic changes in the cell. They found that cells radiated with alpha-particles occasionally show a genetic instability such that a descendent of the damaged cell may at any time in the future suffer a genetic change that then affects all its progeny (but not the other descendents of the originally damaged cell). This type of effect makes the concept of RBE difficult to apply; it would clearly increase the effective RBE for genetic change by alpha-radiation in a very complicated way that would depend on the type of tissue and many other factors. If confirmed, this finding suggests a possible greatly increased risk of late induction of a cancerous change, above that calculated by traditional methods. Their work was with alpha-rays, but the effect may also occur with neutron irradiation.

Nuclear Fission

Some heavy nuclei will split into two smaller nuclei when hit by a neutron with appropriate energy. This is called "fission." The heavy elements have a higher ratio of neutrons to protons than elements in the middle range of weights. Consequently, when a neutron causes a heavy nucleus to fission, more neutrons are often liberated. If one of these happens to hit another of the heavy nuclei, and happens to have the right energy, then that one will fission, and then the next, and the next. If on the average more than one neutron per fission causes another fission, then the reaction will continue and speed up as more and more heavy nuclei fission at the same time. That is a "chain reaction." A huge amount of energy is released in this process.

A nuclear reactor is designed so that the fission reaction is controlled and kept going at a steady speed. Its heat can be used to generate steam, and the excess neutrons to make new isotopes, including plutonium and also several isotopes used in research and in medicine.

A nuclear explosive or bomb is designed so that the reaction speeds up very rapidly to cause a huge explosion. The explosion is much faster than a chemical "High Explosive." A well-designed bomb explodes in under a millionth of a second.

Whether the reaction started by a burst of neutrons does or does not cause a chain reaction depends upon the energy of the neutrons liberated, on the geometry, on the amount of the fissile element present, and on the presence of other substances (called moderators) whose nuclei may either capture neutrons or slow them down.

Two isotopes are suitable for making bombs, uranium-235 and plutonium-239. They are among the few long-lived isotopes capable of sustaining a chain reaction that are possible to obtain in sufficient quantity.

Nuclear reactors are designed so that nuclear fission reactions are controlled and take place much more slowly than in a bomb. One method of control is to include movable rods containing an element like boron that absorbs neutrons without fission. Rods can be withdrawn gradually until the chain reaction starts, and then partially re-inserted to allow fission go on at the desired rate of heat production.

Nuclear Fusion

Another source of energy from atomic nuclei is the fusion of two small nuclei, such as isotopes of hydrogen or lithium, to form a larger nucleus. An even larger amount of energy per gram of reacting substance is liberated in this type of reaction than in nuclear fission. To start fusion a very high temperature is needed, comparable to those in the sun. Such temperatures are produced very briefly during nuclear explosions. They are extremely difficult to produce by any other method. The technology to make sustained, controlled fusion reactors for energy production has not yet been demonstrated.

The "hydrogen bomb" works by fusion. Part of the energy of a fission explosion is focused on a small mass of the hydrogen isotope deuterium in the form of lithium hydride, which also generates the other hydrogen isotope tritium under bombardment with neutrons. Some tritium as such is also present to start the process. Deuterium (H-2) and tritium (H-3) undergo fusion with the formation of helium, neutrons, and an explosive release of energy. The power of the explosion is many times that from a fission bomb of similar size.

Glossary

See the Appendix for fuller explanations of some of these concepts. Unit prefixes and abbreviations/acronyms are listed separately at the end of this section.

Alpha radiation: Radiation consisting of helium nuclei (atomic wt. 4, atomic number 2) that are discharged by radioactive disintegration of some heavy elements, including uranium-238, radium-226, and plutonium-239.

Atomic number: The atomic number of an element is the number of protons in the nucleus of each atom. It determines the chemical properties of the element.

Atomic weight: The nominal atomic weight of an isotope is given by the sum of the number of neutrons and protons in each nucleus. The exact atomic weight differs fractionally from that whole number, because neutrons are slightly heavier than protons and the mass of the nucleus is also affected by the binding energy.

Becquerel: A unit of radioactivity equal to one disintegration per second. It is an extremely small unit, equal to about 27 picocuries.

Beta radiation: Radiation consisting of electrons or positrons emitted in many radioactive disintegrations, at speeds approaching the speed of light.

Calorie: A unit of heat or energy sufficient to raise the temperature of 1 gram of water by 1 degree Celsius. In dietetics, the kilocalorie is the unit usually used, frequently called a "calorie," omitting the prefix.

Critical mass: The amount of a fissile substance that will allow a self-sustaining chain reaction. The amount depends both on the properties of the fissile element and on the shape of the mass.

Curie: The traditional unit of radioactivity equal to the radioactivity of 1 gram of pure radium. It is equal to 37,000,000,000 disintegrations per second (37 billion becquerels).

Decay-correction: The amount by which the calculated radioactivity (for example, of a release of radioisotopes) must be reduced after a period of time, to allow for its radioactive decay during that time.

Electron: An elementary particle carrying 1 unit of negative electric charge. Its mass is 1/1837 that of a proton.

External radiation dose: The dose from sources of radiation located outside the body. This is most often from gamma rays, though beta rays can contribute to dose in the skin and other relatively superficial tissues.

Fission: The splitting of the nucleus of an element into fragments. Heavy elements such as uranium or plutonium release energy when fissioned.

Fission product: Any atom created by the fission of a heavy element. Fission products are usually radioactive.

Fusion: The combining of two nuclei to form a heavier one. Fusion of the isotopes of light elements such as hydrogen or lithium gives a large release of energy.

Gamma radiation: Electromagnetic radiation of high photon energy. The term is used for radiation that comes from radioactive disintegration. X-rays are identical with the lowest energy gamma rays and have sufficient energy to ionize atoms with which they interact.

Gray: A unit of absorbed radiation dose equal to 100 rads.

Half-life: The time in which half the atoms of a radioactive substance will have disintegrated, leaving half the original amount. Half of the residue will disintegrate in another equal period of time.

Induced radioactivity: Radioactivity produced in any material as a result of nuclear reactions, especially by absorption of neutrons.

Internal radiation dose: The dose to organs of the body from radioactive material inside the body. It may consist of any combination of alpha, beta, and gamma radiation.

Ionize: To split off one or more electrons from an atom, thus leaving it with a positive electric charge. The electrons usually attach to other atoms or molecules, giving them a negative charge.

Isotope: The atoms of any one element all have the same number of protons (and hence the same chemical properties) but may have different numbers of neutrons and, therefore, different weights. Thus, there is more than one kind of atom for any one element, and the different kinds are called the isotopes of that element. Some isotopes are stable; others are unstable and therefore radioactive (radioisotopes).

Kiloton (KT): In the context of nuclear weapons, this term, which means 1,000 tons, is always used as a measure of explosive power. It is equal to the explosive power of 1,000 tons of TNT.

Micron: One one-millionth of a meter.

Neutron: An elementary particle slightly heavier than a proton, with no electric charge.

Nucleus: The nucleus of an atom is the central core that comprises almost all the weight of the atom. All atomic nuclei (except H-1, which has a single proton) contain both protons and neutrons.

Photon: The indivisible unit, or quantum, of electro-magnetic radiation. The energy of the photons determines the nature of the radiation, from radio waves at the lowest energy levels, up through infra-red, visible, and ultra-violet light, to X- or gamma-rays, which have energy high enough to ionize atoms.

Positron: An elementary particle with a positive electric charge, but in other respects identical with an electron.

Proton: An elementary particle with a positive electric charge and a mass that is given the value 1 on the scale of atomic weights.

Rad: A unit of absorbed dose of radiation defined as deposition of 100 ergs of energy per gram of tissue. It amounts to approximately one ionization per cubic micron.

Radioactivity: The spontaneous discharge of radiation from atomic nuclei. This is usually in the form of beta or alpha radiation, together with gamma radiation. Beta or alpha emission results in transformation of the atom into a different element, changing the atomic number by +1 or -2 respectively.

Radionuclide: Any radioactive isotope.

Relative Biological Effectiveness (RBE): A factor that can be determined for different types of ionizing radiation, representing the relative amount of biological change caused by 1 rad. It depends upon the density of ionization along the tracks of the ionizing particles, being highest for the heavy particles: alpha rays and neutrons.

Rem: A unit of equivalent absorbed dose of radiation, taking account of the relative biological effectiveness of the particular radiation. The dose in rems is the dose in rads multiplied by the RBE.

Roentgen: A unit of gamma radiation measured by the amount of ionization in air. In non-bony biological tissue 1 roentgen delivers a dose approximately equal to 1 rad.

Sievert: A unit of equivalent absorbed dose equal to 100 rems.

Thermonuclear weapon: A nuclear weapon that gets a large part of its explosive power from fusion reactions.

Ton: A metric ton is 1,000 kilograms. This is approximately 2,200 pounds, and very nearly equal to a British ton (2,240 pounds). The U.S. ton is 2,000 pounds. In this book, "ton" means "metric ton" and is used interchangeably with it.

TNT equivalent: The weight of TNT which would release the same amount of energy as a particular nuclear explosion. One ton of TNT releases approximately 1.2 billion calories (that is, 5.1 kiloJoules per gram). Nuclear explosions are usually measured in kilotons (KT) or megatons (MT).

Yield: The energy released by a nuclear explosion.

Unit Prefixes

It is convenient to use a range of standard prefixes to denote large multiples and small fractions of the various units of measurement such as grams, metres, rads, rems, curies or becquerels. The commonly used prefixes are:

tera-	one trillion times	**milli-**	one one-thousandth
giga-	one billion times	**micro-**	one one-millionth
mega-	one million times	**nano-**	one one-billionth
kilo-	one thousand times	**pico-**	one one-trillionth

Note that each of these is one one-thousandth of the preceding one. Also note that one one-millionth of a metre, which would be 1 "micro-metre," is shortened to "1 micron."

Abbreviations and Acronyms

AEC	(U.S.) Atomic Energy Commission
AGR	advanced gas-cooled reactor
BARC	(Indian) Bhabha Atomic Research Center
BNFL	British Nuclear Fuels
Bq	becquerels
C	celsius (or carbon)
cal.	calorie
Ci	curies
CIA	(U.S.) Central Intelligence Agency
DOE	(U.S.) Deparment of Energy
FOIA	(U.S.) Freedom of Information Act
ft.	foot, feet
G	giga-
g.	gram
H	hydrogen
HAST	high-activity storage tank(s)
IAEA	International Atomic Energy Agency
ICPP	(U.S.) Idaho Chemical Processing Plant
INEL	(U.S.) Idaho National Engineering Laboratory
K or k	kilo-
kg.	kilogram
LET	linear energy transfer
M	mega-
m.	meter
MOx	mixed oxide fuel
MW	megawatts
MWt	megawatts thermal
n	nano-
NAS	(U.S.) National Academy of Sciences
NRC	(U.S.) Nuclear Regulatory Commission
O	oxygen
PCM	plutonium-contaminated materials
Pu	plutonium
Purex	plutonium-uranium extraction (process)
NWCF	(U.S.) New Waste Calcining Facility
RBE	relative biological efficiency
Redox	reduction-oxidation (process)
sec.	second
THORP	(U.K.) Thermal Oxide Reprocessing Plant
TRU	transuranic (radioactive waste)
U	uranium
W	watt

References

Ahearne, J. (Chairman, Advisory Committee on Nuclear Facility Safety). 1990. Hanford letter-report to James D. Watkins, Secretary of Energy, July 23.

Albright, D. 1987. Civilian inventories of plutonium and highly enriched uranium. In *Preventing Nuclear Terrorism*, ed. by P. Leventhal and Y. Alexander, pp. 265–291. Cambridge: Lexington Books.

Albright, D. 1988. Reprocessing and enrichment programs in Argentina, Brazil, India, Israel, Pakistan, and South Africa: nuclear explosive materials production. Working Paper, Federation of American Scientists, Washington, D.C. (October.)

Albright, D., and H.A. Feiveson. 1988. Plutonium recycling and the problem of nuclear proliferation. *Annual Review of Energy*, pp. 239–265.

Albright, D., and T. Zamora. 1989. India, Pakistan's nuclear weapons: all the pieces in place. *Bulletin of the Atomic Scientists*, vol. 45, no. 5, pp. 20–26.

Alvarez, R., and A. Makhijani. 1988. Radioactive waste: hidden legacy of the arms race. *Technology Review*, vol. 91, no. 6, pp. 42–51.

Bair, W.J. 1975. Biomedical aspects of plutonium. In *Public Issues of Nuclear Power*, ed. by H.S. Isbin, pp. 686–729. Minneapolis: Univ. of Minnesota.

Bair, W.J., and R.C. Thompson. 1974. Plutonium: biomedical research. *Science*, vol. 183, pp. 715–722.

Barnaby, F., ed. 1991. *Plutonium and Security*. London: Macmillan.

Barrillot, B. 1991. The manufacture of nuclear weapons in France. Documentation and Research Centre for Peace and Conflicts, Lyon, France.

Bates, J.K. 1990. The role of surface layers in glass leaching performance. Presentation to the Materials Research Society conference, Boston, November 26–29.

Bates, J.K., et al. 1982. Hydration aging of nuclear wastes. *Science*, vol. 218, pp. 51–53.

Beard, S.J., P. Hatch, G. Jensen, and E.C. Watson, Jr. 1967. Purex TK-105-A waste storage tank liner instability and its implications on waste containment and control. ARH-78, Atlantic Richfield Hanford. (October 31.)

Benedict, M., T.H. Pigford, and H.W. Levi. 1981. *Nuclear Chemical Engineering*, 2nd edition. New York: McGraw-Hill.

Berkhout, F., and W. Walker. 1990. THORP and the economics of reprocessing. Science Policy Research Unit, University of Sussex, Sussex, U.K. (November.)

Berkhout, F., and W. Walker. 1991a. Spent fuel and plutonium policies in Western Europe. *Energy Policy*, July/August, pp. 553–566.

Berkhout, F., and W. Walker. 1991b. Are current back-end policies sustainable? Presentation to conference "The Management of Spent Nuclear Fuel," London, April 29–30.

Berkhout, F., T. Suzuki, and W. Walker. 1990. The approaching plutonium surplus: a Japanese/European predicament. *International Affairs*, vol. 66, no. 3, pp. 523–543.

Bernstein, B.J. 1985. Radiological warfare: the path not taken. *Bulletin of the Atomic Scientists*, vol. 41, August, pp. 44–49.

Bloomster, C.H., P.L. Hendrickson, M.H. Killinger, and B.J. Jonas. 1990. Options and regulatory issues related to disposition of fissile materials from arms reduction. Prepared for U.S. Department of Energy, PNL-SA-18728, Pacific Northwest Laboratory, Richland, Washington. (December.)

Blush, S.M. (Director, Office of Nuclear Safety, U.S. Department of Energy). 1990. Statement before the U.S. Senate Committee on Governmental Affairs, July 31.

Bolsunovsky, A. 1992. Russian nuclear materials production and environmental pollution. Presentation to the CIS Non-Proliferation Project conference "The Nuclear Predicament in the FSU," Monterey, California, April 6–9.

Brinkley, J. 1991. Israeli nuclear arsenal exceeds earlier estimates, book reports — Soviet and Arab foes are named as top targets. *New York Times*, October 20, p. 1.

Broad, W. 1992. Nuclear accords bring new fears on arms disposal. *New York Times*, July 6, p. 1.

Buldakov, L.A. 1989. Interviewed by Arjun Makhijani (IEER), Moscow, December 7. Interpreting provided by Soviet Committee of Physicians for the Prevention of Nuclear War.

Buldakov, L.A. 1991. Interviewed by Katherine Yih (IPPNW), Moscow, October 31. Interpreting provided by Soviet Committee of Physicians for the Prevention of Nuclear War.

Buldakov, L.A., S.N. Demin, V.A. Kostyucenko, et al. 1989. Medical consequences of the radiation accident in the southern Urals in 1957. Translated from Russian, 89-12474 (5641e/577e), IAEA, Vienna.

Burger, L.L. 1984. Complexant stability investigation, Task 1 — ferrocyanide solids. PNL-5441, Pacific Northwest Laboratories. (November.)

Carter, L.J. 1987. Nuclear imperatives and public trust: dealing with radioactive waste. Resources for the Future, Washington, D.C.

Cash, R.J., and G.B. Mellinger. 1991. Hanford Ferrocyanide Task Team. Integrated program plan for stability of Hanford tanks containing ferrocyanide wastes (pre-decisional draft). WHC-EP-0399 (draft), Westinghouse Hanford Company. (January 10.)

Cochran, T.B., W.M. Arkin, R.S. Norris, and M.M. Hoenig. 1987a. *Nuclear Weapons Databook; Vol. II: U.S. Nuclear Warhead Production.* Cambridge: Ballinger Publishing Co.

Cochran, T.B., W.M. Arkin, R.S. Norris, and M.M. Hoenig. 1987b. *Nuclear Weapons Databook; Vol. III: U.S. Nuclear Warhead Facility Profiles.* Cambridge: Ballinger Publishing Co.

Cochran, T.B., and R.S. Norris. 1992. Russian/Soviet nuclear warhead production. Nuclear Weapons Databook Working Papers, NWD 92-4 (4th revision), Natural Resources Defense Council, Washington, D.C. (June 12.)

Cogema 1989. Cogema: La Compagnie du Cycle du Combustible Nucléaire. Cogema, 78141 Velizy-Villacoublay, France. (May.)

Cogema undated-a. Le Cycle du Combustible Nucléaire. Data sheets on Cogema's reprocessing operations. Cogema, 78141 Velizy-Villacoublay, France.

Cogema undated-b. Retraitement. Booklet on Cogema's reprocessing operations. Cogema, 78141 Velizy-Villacoublay, France.

Cohn, C. 1987. Click'ems, glick'ems, Christmas trees, and cookie cutters: nuclear language and how we learned to pat the bomb. *Bulletin of the Atomic Scientists*, vol. 43, June, pp. 17–24.

Commissariat à l'Energie Atomique. 1988. Outlook on breeders. A special report accompanying *Nucleonics Week*, vol. 24, no. 16 (April 28).

Conant, J.B. 1943. Preliminary statement concerning the probability of the use of radioactive material in warfare. Manhattan Project Archives, Modern Military Branch, National Archives, Washington, D.C. (July 1.)

Coyle, D., L. Finaldi, E. Greenfield, M. Hamilton, E. Hedemann, W. McDonnell, M. Resnikoff, J. Scarlott, and J. Tichenor. 1988. Deadly defense: military radioactive landfills. Radioactive Waste Campaign, New York.

CRC. 1988. *CRC Handbook of Chemistry and Physics*, 69th edition, ed. by R.C. Weast. Boca Raton, Florida: CRC Press, Inc.

Cuddihy, R.G., R.O. McClellan, J. A. Mewhinney, and B.A. Muggenburg. 1976. Correlations between the metabolic behavior of inhaled and intravenously injected plutonium in beagle dogs. In *Health Effects of Plutonium and Radium*, ed. by S.S. Webster, pp. 169–182. Salt Lake City: University of Utah.

Dagle, G.E., R.W. Bristline, and J.L. Level. 1984. Pu induced wounds in beagles. *Health Physics*, vol. 47, pp. 73–84.

Du Pont. 1988. E.I. Du Pont de Nemours & Co. safety analysis — 200 area Savannah River Plant, liquid radioactive waste handling facilities. DPSTSA-200-10, Sup-18, Aiken, S. Carolina. (August.)

Durbin, P.W., and N. Jeung. 1976. Reassessment of distribution of plutonium in the human body based on experiments with non-human primates. In *Health Effects of Plutonium and Radium*, ed. by S.S. Webster, pp. 297–313. Salt Lake City: University of Utah.

Ege, K., and A. Makhijani. 1982. U.S. nuclear threats: a documentary history. *CounterSpy*, vol. 6, no. 4, pp. 8–23.

Eisenbud, Merril. 1987. *Environmental Radioactivity from Natural, Industrial and Military Sources*, 3rd edition. San Diego: Academic Press.

Energy Research and Development Administration (ERDA). 1977. Waste management operations, Savannah River Plant, Aiken, SC — final environmental impact statement. ERDA-1537. (September.)

Evans, H.J. 1992. Alpha-particle after effects. *Nature*, vol. 355, no. 6362, pp. 674–5.

Falci, Frank P. 1990. Final trip report, travel to USSR for fact finding discussions on environmental restoration and waste management, June 15–28, 1990. Office of Technology Development, U.S. DOE.

Federation of American Scientists. 1991a. Ending the production of fissile material for weapons, verifying the dismantlement of nuclear warheads. Washington, D.C. (June.)

Federation of American Scientists. 1991b. Report on Moscow meetings and workshop, December 16–20, 1991. Washington, D.C. (December 30.)

Feiveson, H.A. 1989. *Plutonium Fuel: An Assessment*. Paris: Organization for Economic Cooperation and Development (OECD).

Feiveson, H.A. 1992. What should we do with separated plutonium? Report prepared for Federation of American Scientists/Natural Resources Defense Council, Fourth International Workshop on Nuclear Warhead Elimination, Washington, D.C., February 26–27.

Fieldhouse, R.W. 1991. Chinese nuclear weapons: a current and historical overview. Nuclear Weapons Databook Working Paper, Natural Resources Defense Council, Washington, D.C. (March.)

Ford, D. 1982. *The Cult of the Atom*. New York: Simon and Schuster.

Fujita et al. 1989. *Journal of Radiation Research*, vol. 30, pp. 359–381.

Gerton, R.E. 1990. Hanford waste tank safety concerns. U.S. Department of Energy presentation to the Advisory Committee on Nuclear Facility Safety, May 22.

Gillete, R. 1973. Radiation spill at Hanford: the anatomy of an accident. *Science*, vol. 181, no. 4101 (August 24).

Gofman, J.W. 1981. *Radiation and Human Health*. San Francisco: Sierra Club Books.

Gorbachev Commission. 1991. Proceedings of the Commission to Study the Ecological Situation in Chelyabinsk Oblast; Volume II: The ecological characteristics of Chelyabinsk Oblast; Opinion of the working group. Commission created by order of the USSR President, RP #1283, January 3, 1991.

Grygiel, M. 1991. Plans for disposal of tank waste. Westinghouse Hanford Company presentation at Hanford Technical Exchange Program, November 13.

Hamilton, J.G. 1949. The metabolism of the radioactive elements created by nuclear fission. *New England Journal of Medicine*, vol. 240, no. 22, pp. 863–870.

Hanford Westinghouse. 1952. A study of the effect of pH of first cycle bismuth phosphate waste on the corrosion of mild steel. HW 26202. (November 13.)

Hebel, L.C., et al. 1978. Report to the American Physical Society by the Study Group on Nuclear Fuel Cycle and Waste. *Reviews of Modern Physics*, January.

Hersh, S.M. 1991. *The Samson Option*. New York: Random House.

Hesketh, R.V. 1984. Nuclear power UK, nuclear weapons USA. Evidence on behalf of CND in the Sizewell B Inquiry, September. Published in the Sizewell B public inquiry documents (U.K.)

Heung, L.K. 1982. Gaseous safety in a confined laboratory, users guide. DPST-82-717, as cited in Du Pont 1988.

Hibbs, M. 1988. U.S. repeatedly warned Germany on nuclear exports in Pakistan. German firm's exports raise concern about Pakistan's nuclear capabilities. *Nuclear Fuel*, vol. 13, no. 5 (March 7).

Hollister, C.D., R. Anderson, and G.R. Heath. 1981. Subseabed disposal of nuclear waste. *Science*, vol. 213, no. 4514 (September 18).

International Atomic Energy Agency. 1986. Summary report on the post-accident review meeting on the Chernobyl accident. Safety Series No. 75, INSAG-1, IAEA, Vienna.

International Commission on Radiological Protection (ICRP). 1972. *The Metabolism of Compounds of Pu and other Actinides*. ICRP Publication 19. New York: Pergamon Press.

International Commission on Radiological Protection (ICRP). 1983. *Radionuclide Transformations: Energy and Intensity of Emissions*. ICRP Publication 38, vol. 11–13. Oxford: Pergamon Press.

International Energy Agency. 1989. International Energy Agency government R&D budgets for nuclear breeder programs for 21 countries, 1978–1989. In *Energy Policies and Programmes of IEA Countries*. Paris: International Energy Agency.

International Physicians for the Prevention of Nuclear War and Institute for Energy and Environmental Research. 1991. *Radioactive Heaven and Earth: The Health and Environmental Effects of Nuclear Weapons Testing in, on, and above the Earth*. New York: Apex Press; London: Zed Books.

Jee, W.S.S., and J.S. Arnold. 1960. Radioisotopes in the teeth of dogs — I. The distribution of plutonium, radium, radiothorium, mesothorium and strontium and the sequence of histopathologic changes in teeth containing plutonium. *Arch. Oral Biol.*, vol. 2, pp. 215–38.

Kadhim, M.A., D.A. Macdonald, D.T. Goodhead, S.A. Lorimore, S.J. Marsden, and E.G. Wright. 1992. Transmission of chromosomal instability after plutonium [alpha]-particle irradiation. *Nature*, vol. 355, no. 6362, pp. 738–740.

Keeny, S.M., and W.K.H. Panofsky. 1992. Warhead control regimes. *Arms Control Today*, January/February.

Kossenko, M.M. 1992a. Radiation incidents in the Southern Urals. Urals Research Center for Radiation Medicine, Chelyabinsk, Russia.

Kossenko, M.M. 1992b. Medical effects of population irradiation on the River Techa: radiation risk assessment. Urals Research Center for Radiation Medicine, Chelyabinsk, Russia.

Kossenko, M.M., M.O. Degteva, and M.A. Petrushova. 1992. Estimate of leukemia risk to those exposed as a result of nuclear incidents in the southern Urals. *PSR Quarterly*, vol. 2, no. 4 (forthcoming).

La Gazette Nucléaire. 1989. No. 98/99 (December). (Published in Orsay, France.)

Langham, W.H., S.H. Bassett, P.S. Harris, and R.E. Carter. 1950. Distribution and excretion of plutonium administered intraveneously to man. Joint report from the Los Alamos Scientific Laboratory of the University of California and the Atomic Energy Project of the University of Rochester School of Medicine and Dentistry. LA—1151, Copy 35 of 50, Series A. (September 20.)

Large, J. 1992a. Letter to Arjun Makhijani re: high level waste storage. Large and Associates, London, January 6.

Large, J. 1992b. Written comments to IEER on manuscript of *Plutonium: Deadly Gold of the Nuclear Age*. Large and Associates, London, April 21.

Lenssen, N. 1991. Nuclear waste: the problem that won't go away. Worldwatch Paper 106, Worldwatch Institute, Washington, D.C.

Lewis, John Wilson, and Xue Litai. 1988. *China Builds the Bomb*. Stanford University Press.

Lipshutz, R.D. 1980. *Radioactive Waste: Politics, Technology and Risk. A report of the Union of Concerned Scientists*. Cambridge, Massachusetts: Ballinger Publishing Co.

Makhijani, A. 1989. Reducing the risks: policies for the management of highly radioactive nuclear waste. Institute for Energy and Environmental Research, Takoma Park, Maryland.

Makhijani, A. 1991. Glass in the rocks: some issues concerning the disposal of radioactive borosilicate glass in a Yucca Mountain repository. Prepared for the Nevada Nuclear Waste Task Force, January 29.

Makhijani, A., and M. Hoenig. 1991c. A black market in red nukes. *The Washington Post*, September 29, p. C1.

Makhijani, A., and S. Saleska. 1992. *High-Level Dollars, Low-Level Sense: A Critique of Present Policy for the Management of Long-Lived Radioactive Waste and Discussion of an Alternative Approach*. New York: The Apex Press.

Makhijani, A., and K. Tucker. 1985. Heat, high water, and rock instability at Hanford: a preliminary assessment of the suitability of the Hanford, Washington site for a high-level nuclear waste repository. Health and Energy Institute, Washington, D.C.

Makhijani, A., R. Alvarez, and B. Blackwelder. 1986. Deadly crop in the tank farm: an assessment of the management of high-level radioactive wastes in the Savannah River Plant tank farm. Based on official documents. Environmental Policy Institute, Washington, D.C.

Makhijani, A., R. Alvarez, and B. Blackwelder. 1987. Evading the deadly issues: an assessment of the Du Pont Corporation's management of high-level nuclear wastes at the federal Savannah River Plant. Environmental Policy Institute, Washington, D.C.:

Makhijani, A., S. Saleska, and M. Ospina. 1991a. Sources of fire and explosion risk in high-level nuclear waste storage tanks at U.S. Department of Energy nuclear weapons production sites. Presentation to conference "The Environmental Consequences of Nuclear Development", University of California at Irvine, April 11–14.

Mark, J.C., et al. 1987. Can terrorists build nuclear weapons? In _Preventing Nuclear Terrorism_, ed. by P. Leventhal and Y. Alexander. Cambridge: Lexington Books.

May, J. 1989. _The Greenpeace Book of the Nuclear Age: The Hidden History, The Human Cost_. London: Victor Gollancz Ltd.

McClellan, R.O. 1979. Annual report of the Inhalation Toxicology Research Institute operated for the United States Department of Energy by the Lovelace Biomedical and Environmental Research Institute, Inc. LF-69: UC-48. (December.)

Medvedev, Zhores Aleksandrovich. 1976. _New Scientist_, November 4, p. 264.

Medvedev, Zhores Aleksandrovich. 1977. _New Scientist_, June 30, pp. 761–764.

Monroe, S.D. 1991. Chelyabinsk: the evolution of disaster. _Soviet Environmental Watch_ (Monterey Institute of International Studies, Monterey, California), no. 1 (fall), pp. 18–29.

Natural Resources Defense Council, Hanford Education Action League, and Nuclear Safety Campaign. 1990. Notice of intent to sue. Letter to U.S. Department of Energy Secretary Watkins requesting the preparation of a supplemental environmental impact statement prior to Purex restart, January 12. Available from NRDC, 1350 New York Ave. NW, Washington, D.C. 20005.

Nazarov, A.G., E.B. Burlakova, D.P. Osanov, G.S. Sakylin, L.H. Shadrin, B.A. Whevchenko, E.A. Yakovkeva, I.A. Seleznev, N.I. Mironova, K.V. Kyranov, and I.I. Pavlinova. 1991. Resonance: southern Urals atom: to be or not to be? Permanent Expert Group of the Supreme Soviet of the USSR; A.N. Penyagin, ed. Chelyabinsk, Russian Federation: Southern Urals Book Publishers.

New York Times. 1990. New explosion threat seen at nuclear plant. March 25, p. 31.

Nikipelov, B.V. 1989. Experience in managing the radiological and radioecological consequences of the accidental release of radioactivity which occurred in the Southern Urals in 1957. Translated from Russian, 89-12472 (5626e/586), IAEA, Vienna.

Nikipelov, B.V., and E.G. Drozhko. 1990. An explosion in the Southern Urals. _Priroda_, no. 5 (897). English translation for Lawrence Livermore National Laboratory, Livermore, California, pp. 4–11.

Nuclear Fuel. 1992. January 26.

Nuclear Safety. 1960. Plutonium release from the Thorex pilot plant. _Nuclear Safety_, no. 1, pp. 78–80.

Nuke Info Tokyo. 1990. No. 16 (March–April). (Newsletter of Citizens' Nuclear Information Center, Tokyo.)

Odell, M. 1992. Vitrification—world review. _Nuclear Engineering International_, June, pp. 51–52.

Park, J.F., W.J. Bair, and R.H. Busch. 1972. Progress in beagle dog studies with transuranium elements at Battelle-Northwest. *Health Physics*, vol. 22, pp. 803–10.

Parker, F.L. 1978. Analysis of the Medvedev report of a Soviet radioactive waste incident. Vanderbilt University, Nashville, Tennessee.

Peden, W. 1991. Britain's bomb (draft). Commissioned for Greenpeace UK. (April.)

Penyagin, A. (former USSR Supreme Soviet Committee on Environment and Management of Natural Resources, former Chairman of Subcommmittee on Atomic Power and Nuclear Ecology). 1991. Interviewed by K. Yih (IPPNW) and A. Brooks, Moscow, October 30 and November 6.

Perry, J.H., et al. 1963. *Chemical Engineers' Handbook*, 4th edition.

Petrovsky, V.F. 1989. Statement to the U.N. General Assembly, New York, October 25.

Physicians for Social Responsibility. 1992. *Dead Reckoning: A Critical Review of the Department of Energy's Epidemiologic Research*. A report by PSR's Physicians Task Force on the Health Risks of Nuclear Weapons Production. Washington, D.C.: Physicians for Social Responsibility.

Romanov, G.N., and A.S. Voronnov. 1990. Radiation situation after the accident. *Priroda*, no. 5 (897). English translation for Lawrence Livermore National Laboratory, Livermore, California, pp. 11–19.

Rotblat, J. 1992. Written comments to IPPNW on manuscript of *Plutonium: Deadly Gold of the Nuclear Age*. Pugwash Conferences on Science and World Affairs, London, June 1.

Rowen, Henry. 1991. Risk assessment and organizational behavior: the case of nuclear technology and the spread of nuclear weapons. In *Risks, Organizations, and Society*, ed. by M. Shubick. Boston: Kluwer Academic Publishers.

Saleska, S. 1991. Notes from meeting of Technical Advisory Panel of U.S. Department of Energy, held in Kennewick, Washington, November 13–14.

Saleska, S. 1992a. U.S. radiation protection standards for the general public. IEER Activist Paper, no. 1 (February). Institute for Energy and Environmental Research, Takoma Park, Maryland.

Saleska, S. 1992b. New evidence on low-dose radiation exposure. *Science for Democratic Action* (publication of Institute for Energy and Environmental Research, Takoma Park, Maryland), vol. 1, no. 2, pp. 1–2, 4.

Saleska, S. 1992c. Ecological consequences of nuclear weapons development in the southern Urals: a conference and working trip in Chelyabinsk, Russia, May 15–30, 1992. Institute for Energy and Environmental Research, Takoma Park, Maryland.

Saleska, S., and A. Makhijani. 1990. To reprocess or not to reprocess: the Purex question; a preliminary assessment of alternatives for the management of N-reactor irradiated fuel at the U.S. Department of Energy's Hanford nuclear weapons production facility. Prepared for Hanford Action Education League, Spokane, Washington; Institute for Energy and Environmental Research.

Saleska, S., et al. 1989. Nuclear legacy: an overview of the places, problems, and politics of radioactive waste in the United States. A report for Public Citizen's Critical Mass Energy Project, 215 Pennsylvania Ave. SE, Washington, D.C. 20003. (September.)

Sanger, D.E. 1991. Japan's plan to import plutonium arouses fear that fuel would be hijacked. *New York Times*, November 25, p. D8.

Sanger, S.L., with R.W. Mull. 1989. *Hanford and the Bomb, An Oral History of World War II*. Seattle: Living History Press.

Schneider, K. 1992. Nuclear disarmament raises fear on storage of 'triggers'. *New York Times*, February 26, p.1.

Sege, G. 1953. Overconcentration in initial operation of uranium evaporator-231 Building. HW-28690, Richland, Washington.

Snake River Alliance Bulletin. 1992. Cleaning up cleanup. Vol. 6, no. 7 (August). (Newsletter of Snake River Alliance, Boise, Idaho.)

Soran, D., and D.B. Stillman. 1982. An analysis of the alleged Kyshtym disaster. Los Alamos National Laboratory, Los Alamos, New Mexico.

Soyfer, V.N., M.O. Degteva, M.M. Kossenko, A.A. Akleev, V.P. Kozheurov, and G.N. Romanov. 1992. Radiation accidents in the Southern Urals (1949–1967). Laboratory of Molecular Genetics, Department of Biology, George Mason University, Fairfax, Virginia.

Spector, L.S. 1985. *The New Nuclear Nations*. New York: Vintage Books.

Spector, L.S. 1988. *The Undeclared Bomb*. Cambridge: Ballinger Publishing Co.

Spector, L.S., and J.R. Smith. 1990. *Nuclear Ambitions*. Boulder: Westview Press.

Stimson, H.L. 1945. Memo discussed with the President, April 25. Manhattan Project Archives, Modern Military Branch, National Archives, Washington, D.C.

Strode, J.N., et al. 1988. 1988 tank farm waste volume projections. Prepared for the U.S. Department of Energy, Assistant Secretary for Defense Programs; WHC-EP-0197, Westinghouse Hanford Company. (September.)

Sullivan, L.H., S.W. Eisenhawer, J.R. Travis, R.J. Henninger, D.R. MacFarlane, J.R. Coleman, J.W. Spore, T.L. Wilson, B.D. Nichols, and S.C. Hill. 1992. Reanalysis of the safety hazards associated with a hydrogen burn in Hanford high-level waste tank 241-SY-101. Study done by Los Alamos National Laboratory for Tank Analysis Task Force Panel.

Suzuki, T. 1991. Japan's nuclear predicament. *Technology Review*, vol. 94, no. 7, pp. 41–49.

Swinbanks, D. 1991a. Japan debates plutonium. *Nature*, vol. 352, no. 6330 (July 4), p. 7.

Swinbanks, D. 1991b. Japan promotes nuclear power. *Nature*, vol. 353, no. 6347 (October 31), p. 782.

Taylor, T.B. 1990. Dismantlement and fissile material disposal. In *Reversing the Arms Race*, ed. by von Hippel and Sagdeev, pp. 91–115. New York: Gordon and Breach.

Thomas, J. 1990. Shadows of Hanford's past: radiation releases. *Perspective* (newsletter of Hanford Education Action League, Spokane, Washington), vol. 1, no. 3 (fall).

Thomas, J. 1992. Article in *Perspective* (newsletter of Hanford Education Action League, Spokane, Washington), no. 10 (fall).

Thompson, R.C. 1989. Life-span effects of ionizing radiation in beagle dogs: a summary account of four decades of research funded by the U.S. Department of Energy and its predecessor agencies. Pacific Northwest Laboratory, Richland, Washington.

Tomlinson, R.E. 1953. Unusual incident at Savannah River. HW27122, Hanford Works, General Electric Company, Richland, Washington.

Trabalka, J.R., L.D. Eyman, F.L. Parker, E.G. Struxness, and S.I. Auerbach. 1979. Another perspective of the 1958 Soviet nuclear accident. *Nuclear Safety*, vol. 20, no. 2 (March–April).

Trabalka, J.R., L.D. Eyman, and S.I. Auerbach. 1980. Analysis of the 1957–1958 Soviet nuclear accident. *Science*, vol. 209, no. 3354, pp. 345–353.

U.K. Central Electricity Generating Board (CEGB) & South of Scotland Electricity Board (SSEB). 1986. The CEGB/SSEB response to Recommendation 18 in the Environment Committee's report on radioactive waste: an overview.

U.K. Nirex. 1988. The 1987 United Kingdom radioactive waste inventory. Report prepared by Electrowatt Engineering Services for U.K. Nirex and Department of the Environment, DOE-RW-88.061, U.K. Nirex Report no. 54. (October.)

United Nations Scientific Committee on the Effects of Atomic Radiation (UNSCEAR). 1988. *Sources, Effects, and Risks of Ionizing Radiation*. Report to the General Assembly, with annexes. New York: United Nations.

U.S. Central Intelligence Agency. 1959. Accident at the Kasli Atomic Plant. Report no. CS-3/389, 785. (March 4.)

U.S. Central Intelligence Agency. 1961. Miscellaneous information on nuclear installations in the USSR. Report no. C5K-3/465, 141. (February 16.)

U.S. Congress, Office of Technology Assessment. 1986. Staff paper on the subseabed disposal of high-level radioactive waste. Washington, D.C.: U.S. Government Printing Office.

U.S. Congress, Office of Technology Assessment. 1989. The containment of underground nuclear explosions. OTA-ISC-414. Washington, D.C.: U.S. Government Printing Office.

U.S. Department of Energy. 1979. Management of commercially generated radioactive waste. DOE/EIS-0046-D, U.S. DOE, Washington, D.C. (April.)

U.S. Department of Energy. 1982. Draft environmental impact statement: operation of Purex and Uranium Oxide Plant facilities, Hanford Site. DOE/EIS-0089D, U.S. DOE, Washington D.C. (May.)

U.S. Department of Energy. 1984. Spent fuel and radioactive waste inventories, projections, and characteristics. DOE/RW-0006, prepared by Oak Ridge National Laboratory, U.S. DOE, Washington, D.C. (September.)

U.S. Department of Energy. 1986. Draft environmental impact statement: process facility modifications project, Hanford site. DOE/EIS-0115D, U.S. DOE, Washington, D.C. (April.)

U.S. Department of Energy. 1987. Disposal of Hanford defense high-level, transuranic and tank wastes, final environmental impact statement. DOE/EIS 0113, U.S. DOE, Washington, D.C. (December.)

U.S. Department of Energy. 1988. Draft 1988 mission plan amendment. DOE/RW-0187, U.S. DOE, Washington, D.C. (June.)

U.S. Department of Energy. 1989a. Report to Congress on reassessment of the civilian radioactive waste management program. DOE/RW-0247, U.S. DOE, Washington, D.C. (November.)

U.S. Department of Energy. 1989b. Draft supplemental environmental impact statement, Waste Isolation Pilot Plant. DOE/EIS-0026-DS. (April.)

U.S. Department of Energy (Eli Maestas et al.). 1990a. Investigation of potential flammable gas accumulation in Hanford tank 101-SY. Ad-hoc investigation team report to Leo Duffy, Director, Office of Environmental Restoration and Waste Management. (April 4.)

U.S. Department of Energy, Office of Nuclear Safety. 1990b. Report on the handling of safety information concerning flammable gases and ferrocyanide at the Hanford waste tanks. DOE/NS-0001P. (July.)

U.S. Department of Energy. 1990c. Final supplemental environmental impact statement, Waste Isolation Pilot Plant. (January.)

U.S. Department of Energy. 1991a. Independent engineering review of the Hanford waste vitrification system. DOE/EM-0056P, U.S. DOE, Washington, D.C. (October.)

U.S. Department of Energy. 1991b. Integrated data base for 1991: U.S. spent fuel and radioactive waste inventories, projections and characteristics. DOE/RW-0006, Rev. 7, prepared by Oak Ridge National Laboratory for the Office of Civilian Radioactive Waste Management and the Office of Environmental Restoration and Waste Management, U.S. DOE, Washington, D.C.(October.)

U.S. Department of Energy. 1991c. Draft historical perspective of radioactively contaminated liquid and solid wastes discharged or buried in the ground at Hanford. TRAC-0151-VA. (April 5.)

U.S. Environmental Protection Agency. 1985. Environmental radiation protection standards for management and disposal of spent nuclear fuel, high-level and transuranic radioactive wastes; final rule. 40 CRF Part 191. *Federal Register*, vol. 50, no. 1982 (September 19). Washington, D.C.: Government Printing Office.

U.S. Environmental Protection Agency. 1990. Transuranium elements. EPA 520/1-90-015 and -016, Office of Radiation Programs, Washington, D.C. (June.)

U.S. General Accounting Office. 1989. DOE's management of single-shell tanks at Hanford, Washington. GAO/RCED-89-157, GAO, Washington, D.C. (July.)

U.S. Joint Chiefs of Staff. 1947. The evaluation of the atomic bomb as a military weapon. The final report of the Joint Chiefs of Staff Evaluation Board for Operation Crossroads, enclosure to Joint Chiefs of Staff. Document number JCS 1691/7, Record Group RG 218, Modern Military Branch, National Archives, Washington, D.C.

U.S. National Academy of Sciences. 1957. *The Disposal of Radioactive Waste on Land.* Report of the Committee on Waste Disposal of the Division of Earth Sciences. Washington, D.C.: National Academy of Sciences.

U.S. National Academy of Sciences. 1983. *Scientific Basis for Risk Assessment and Management of Uranium Mill Tailings.* Washington, D.C.: National Academy Press.

U.S. National Academy of Sciences. 1989. *The Nuclear Weapons Complex: Management for Health, Safety, and the Environment.* Washington, D.C.: National Academy Press.

U.S. National Academy of Sciences. 1990. *Health Effects of Exposure to Low Levels of Ionizing Radiation* (BEIR V). Washington, D.C.: National Academy Press.

U.S. Nuclear Regulatory Commission. 1991. Standards for protection against radiation; final rule. 10 CFR Part 20. *Federal Register,* vol. 56, no. 98 (May 21), Part IV: Rules and Regulations.

Van Tuyl, H. 1983. Potential for exothermic chemical reactions in waste tanks. Pacific Northwest Laboratories. (February 3.)

Voelz, G.L. 1975. What we have learned about plutonium from health data. *Health Physics,* vol. 29, pp. 554–555.

Voelz, G., J. Umbarger, J. McInroy, and J. Healy. 1976. Considerations in the assessment of plutonium deposition in man. In *Diagnosis and Treatment of Incorporated Radionuclides* (conference proceedings), pp. 163–175. Vienna: International Atomic Energy Association.

Voelz, G.L., L.H. Hempelmann, J.N.P. Lawrence, and W.D. Moss. 1979. A 32-year medical follow-up of Manhattan Project plutonium workers. *Health Physics,* vol. 37, pp. 445–485.

Voelz, G.L., and J.N.P. Lawrence. 1991. A 42-year medical follow-up of Manhattan Project plutonium workers. *Health Physics,* vol. 61, no. 2 (August), pp. 181–190.

von Hippel, F. 1992. Controls on nuclear warheads and materials. Statement prepared for the U.S. Senate Armed Services Committee hearings on the disposition of U.S. and C.I.S. strategic nuclear warheads under the START treaties, August 4.

von Hippel, F., D.H. Albright, and B.G. Levi. 1986. Quantities of fissile materials in US and Soviet nuclear weapons arsenals. PU/CEES Report No. 168, Center for Energy and Environmental Studies, Princeton, New Jersey. (July.)

Wald, M.L. 1986. Report assails safety of nuclear waste storage at Carolina plant. *New York Times,* July 24.

Wald, M.L. 1987. Explosion risk at nuclear site is reported high. *New York Times*, September 18.

Warren, S.L. 1946. Memorandum to Commander Task Group 1.2, USS Haven, August 13. Obtained from the archives of the papers left by Col. Warren to the library of the University of California at Los Angeles.

Weinberg, A. 1972. The safety of nuclear power. Presentation to the Council for the Advancement of Science Writing Briefing on New Horizons in Science, Boulder, Colorado, November 14.

Wilson, C.L. 1979. Nuclear energy: what went wrong? *Bulletin of the Atomic Scientists*, vol. 35, no. 6 (June), p. 15.

Wilson, G.F. 1977. Central Intelligence Agency letter to Mr. Richard E. Pollock, Director, The Citizen's Movement for Safe and Efficient Energy, Critical Mass, Washington D.C., November 11.

WIN. 1990. Aqueous liquid waste management. WIN safety analysis report, Section 4.2.2., #WIN 107-4.2, Rev. 4A (as released by Department of Energy). (June.)

Wodrich, D.D. 1989. NRC concerns about grouting double shell tank waste. Internal memorandum, Westinghouse Hanford Company, March 16.

Wodrich, D.D. 1990. Hanford site tank waste stability. Internal memorandum, Westinghouse Hanford Company, February 8.

Wodrich, D. 1991. Status report on evaluation of the single-shell tank low pH issue. Westinghouse Hanford Company, presented to the High-Level Waste Technical Advisory Panel, November 15.

Womack, J.C. (Hanford tank farm process engineer). 1977. Program for stabilization of Tank 105-A. Internal memorandum, September 16.

Wright, R. 1992. *Stolen Continents, The Americas Through Indian Eyes Since 1492.* Boston: Houghton Mifflin Company.